GUT-BRAIN SECRETS

PART 2:

How Nature Nurtures and Man Manipulates

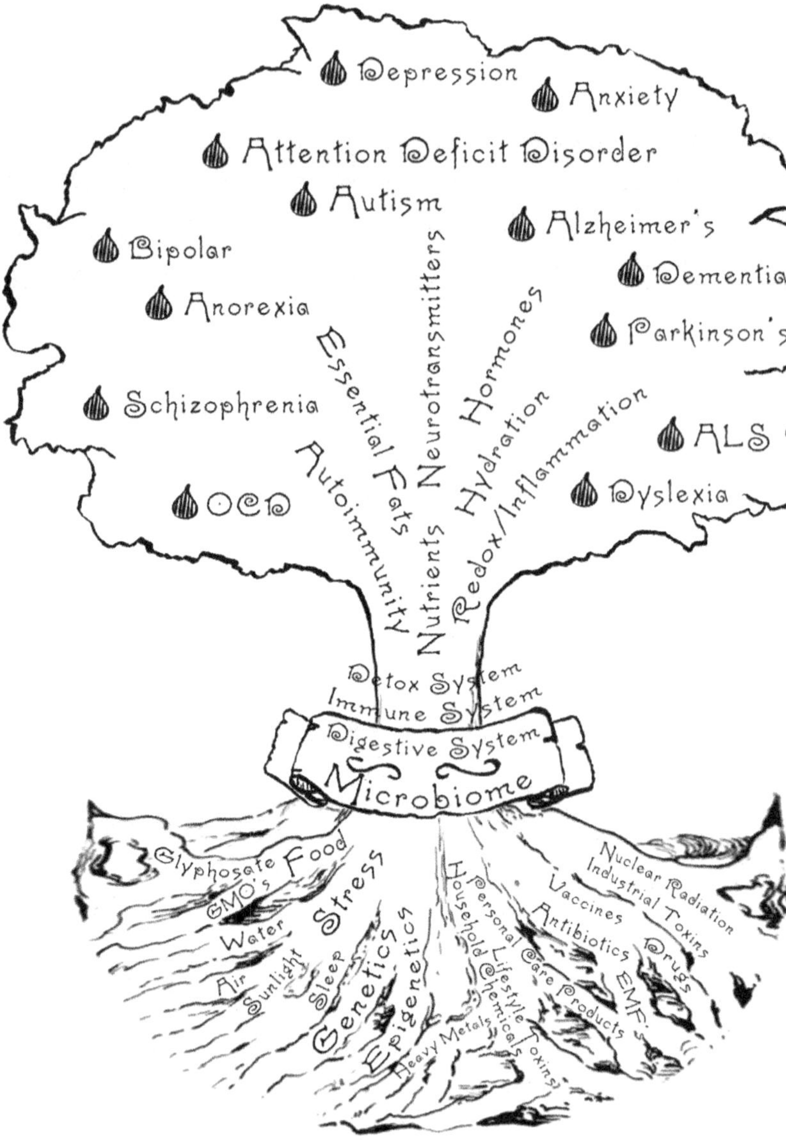

Depression
Anxiety
Attention Deficit Disorder
Autism
Alzheimer's
Bipolar
Dementia
Anorexia
Parkinson's
Schizophrenia
ALS
OCD
Dyslexia
Essential Fats
Neurotransmitters
Hormones
Hydration
Redox/Inflammation
Autoimmunity
Nutrients
Detox System
Immune System
Digestive System
Microbiome
Glyphosate
Food
GMO's
Water
Air
Sunlight
Sleep
Stress
Genetics
Epigenetics
Heavy Metals
Household Chemicals
Personal Care Products
Lifestyle
Toxins
Nuclear Radiation
Industrial Toxins
Vaccines
Antibiotics
Drugs
EMF's

GUT-BRAIN SECRETS

PART 2 (of 6):

How Nature Nurtures and Man Manipulates

(+ How **Redox Signaling Molecules** Æffect **Healing** and **Inflammation**)

R. D. Lee

Gut-Brain Secrets, Part 2: How Nature Nurtures and Man Manipulates (2nd Ed.)

Custom illustrations: Mike Bocianowski.

Cover design and some interior elements: Vincent Vi, www.vincentvi.com.

ISBN-13: 978-1727559026
ISBN-10: 1727559029

Gut-Brain Secrets is available in four formats: (1) As a complete collection, (2) as individual, topic-specific books, (3) in print form, and (4) as electronic ebooks. The all-in-one print book (Parts 1-6) has the benefit of the original formatting, an Index, and pictures as they were meant to be seen. The ebooks have full color images and live hyperlinks.

1. Good Food, Bad Food.
2. How Nature Nurtures and Man Manipulates.
3. Microbes and Our Corrupted Guts.
4. The Science and Mystery of ADD.
5. Detoxification and Elimination.
6. Solutions to Gut-Brain Problems.

Acknowledgements

A big thank you goes to the subject matter experts who generously supported the creation of *Gut-Brain Secrets* by making their information publicly available, and/or enriching the content with their input.

1. Dr. Natasha Campbell-McBride.
2. Dr. Zach Bush.
3. Dr. Elaine Ingham.
4. Dr. Don Huber.
5. Gary Samuelson, PhD.
6. Dr. Christopher Shade.
7. Barbara O'Neill, ND.
8. Kerri Rivera.
9. Dr. Russell Blaylock.
10. Dr. Fereydoon Batmanghelidj

The concepts taught by these subject matter experts also went into this work:

Hippocrates, Sally Fallon Morell, Ken Rohla, Dr. David Getoff, Jeffrey Smith, Gary Null, Donna Gates, Dr. Michael Antoniou, Dr. Derrick MacFabe, Steven Druker, Dr. Thierry Vrain, Clint Ober, Stephanie Seneff, PhD, Sofia Smallstorm, Bruce Lipton, Vaughn Lawrence, Winston Kao, Dr. Eric Berg, Berny Dohrmann and CEO Space, Izabella Wentz, PharmD, Dr. John Bergman, Dr. Gary Kaplan, Dr. Alessio Fasano, Dr. Joseph Mercola, Dr. Hal Huggins, David Hudson, Jim Humble, Dolores Cannon, Ryan Dougherty, Dr. Robert O. Young, Dr. Andreas Kalcker, Brittany Auerbach, Paul Chek, and David Wilcock.

Contents

Disclaimer

THIS IS A FICTIONAL STORY TO BE USED FOR ENTERTAINMENT PURPOSES ONLY. IT'S INTENDED FOR A MATURE AUDIENCE.

IF YOU'RE NEW TO THIS SORT OF INFORMATION, IT MAY SERIOUSLY SHAKE UP YOUR WORLDVIEW. IF YOU HAVE THE SLIGHTEST BIT OF HUMANITY, EMPATHY, AND URGE FOR SELF-PRESERVATION, THIS INFORMATION WILL MAKE YOU UNCOMFORTABLE AND PROBABLY UPSET. IT MAY EVEN CAUSE YOU TO RETHINK EVERYTHING YOU THOUGHT YOU KNEW ABOUT THE NATURE OF YOUR REALITY AND EVERYTHING IN IT. ...FAIR WARNING.

IT IS NOT INTENDED TO BE USED AS A SUBSTITUTE FOR PROFESSIONAL MEDICAL ADVICE, DIAGNOSIS OR TREATMENT. THE AUTHOR(S) MAKE NO REPRESENTATIONS, AND ASSUME NO RESPONSIBILITY, CONCERNING THE ACCURACY, SAFETY, EFFICACY, OR APPROPRIATENESS OF THE PRODUCTS, PROCEDURES, TESTS, TREATMENTS, SERVICES, OPINIONS, OR OTHER INFORMATION CONTAINED IN THE STORY.

IF YOU SUFFER A LIFE THREATENING EMERGENCY WHILE YOU'RE READING THIS BOOK, RESIST THE URGE TO CONTINUE READING. PUT THE BOOK DOWN IMMEDIATELY AND DIAL '911'.

PREFACE:

What *Gut-Brain Secrets* is all About

WARNING: This book contains heretical content

What you're about to read in these pages may come as a shock to you. It's sure to blow your mind, because the principles, real-world clinical experiences, and conclusions contained in this book may cause you to question everything you know – or thought you knew – about how the human body works, how toxins affect the human body, and eating for optimal health.

It may even upset you to find out that the things you've been told about what causes health and disease, from the ground up, have been appallingly misleading, baseless, incomplete, deceptive, or just plain fraudulent.

Reading this book may cause you to think I'm either delusional, or that the concepts presented within actually makes sense and deserve more credit than the carefully crafted rubbish mainstream medicine has been pumping out for decades.

In other words, if you only like to hear what others *think* you want to hear, you're definitely not going to like this story. If you truly believe ignorance is bliss, you might as well quit reading now, because this material takes you on a mind-bending journey light years away from The Average Joe's frame of reference and comfort zone, into how the world *really* works.

If you don't mind having your perspective shifted in a seismic way… That is, if you don't mind having your world-view shattered and rebuilt on a more accurate foundation, then put your "common sense" hat on, because I'm going to take you on a reality-shifting excursion into the real causes of Attention Deficit Disorder and other mental conditions.

I'll point out the root causes of ADD, and what you can do to reverse its physical origins, based on the knowledge, experience, research, and teachings of the world's leading subject matter experts in functional/integrated/alternative medicine – whatever you call healing from within.

Feel free to disagree with their ideas, my explanations, and our conclusions. Because, as you'll soon see, the more we learn about nutrition, disease, and the human body, the more embarrassed conventional medicine becomes at just how wrong it's been … and how right Nature's been from the very start. Keep your mind open and be your own judge.

This material focuses primarily on the physical origins of ADD and GAPS, rather than psychological, behavioral, social, and developmental causes

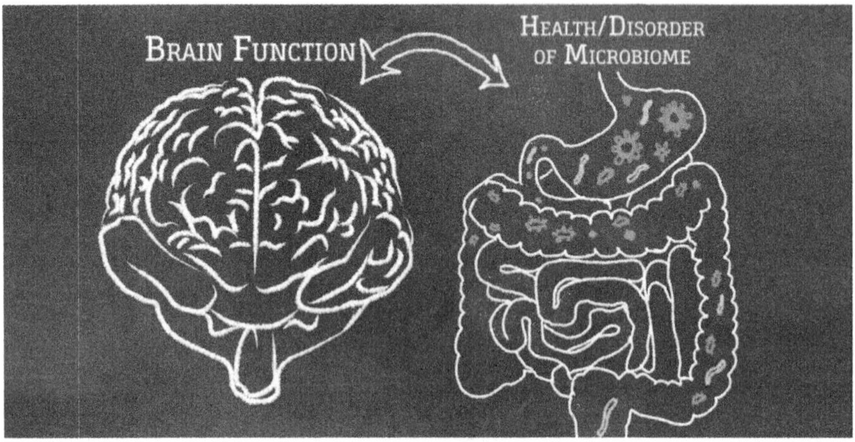

This book's main purpose is to explain how a dysfunctional gut (physical) leads to a dysfunctional brain (psycho-social-developmental). It encompasses the many factors that go into the inception and perpetuation of ADD, as well as the many ways in which the root cause of ADD – a Corrupted Gut – affects the individual biochemically, physically, and immunologically.

A "Corrupted Gut" is my umbrella term to describe unhealthful, imbalanced conditions in the gut. "Gut dysbiosis" is the scientific name for a Corrupted Gut that means the following collection of events occurring together more often than not:

(1) Too many pathogenic (disease-causing) microorganisms in the gut, and too few good, probiotic bacteria (they're opposite sides of the same coin), (2) broken tight junctions causing leaky gut and leaky membranes, (3) improper digestion and mal-absorption, and (4) usually immune system problems.

Simply put, when your gut's a mess, that's a Corrupted Gut. And when your gut falls into a corrupted state, that opens the door for psychological abnormalities to occur. Meaning, disorderly thinking and problematic behavior are *physical* in nature first, *psychological* in origin second or third. Both are symptoms of imbalance and dysfunction happening elsewhere in the body, rather than the real source of what bothers you.

And, as you'll soon find out, I'm big on recognizing the real root of your problems, so you can always focus your efforts on solving real problems, not symptoms and side effects (shocking how often we misunderstand the nature of our problems, and consequently accept partial solutions).

Many patients and practitioners all over the world now refer to these gut-brain conditions as "Gut and Psychology Syndrome" conditions (or GAPS for short), based on Dr. Natasha Campbell-McBride's book and protocol by the same name.

We also explore how the human mind works

We'll uncover many mysteries behind the non-physical aspects of the human mind as we discuss how thoughts, consciousness, and compulsions are formed, as well as how the brain is connected to spirit.

We're going to dissect and examine the mechanisms that cause ADD and GAPS conditions. And by doing that, I hope to give you, your family, your practitioners, and the medical community insight into why ADD individuals think and do things that society deems counter-productive, anti-social, or just plain odd.

Demystified within: The true sources and inner workings of (1) gut dysbiosis, (2) neurological conditions (problems with brain function), (3) mysteries of the mind, (3) immune system dysfunction, as well as (4) chronic, degenerative physical diseases.

By bringing specificity and connections to the ADD conversation, my goal (and by extension, their goal) is to point everyone in the direction of better solutions. Over time, we can reverse the explosive rise in ADD and GAPS, and reduce their occurrence inside the home and out.

This information is collected straight from the cutting edge

As far as mainstream medicine and nutritional sciences are concerned, this information is speculative and unproven. On the other side, as far as holistic healers – the real healers – are concerned, this information is simply ancient wisdom being rediscovered and explained scientifically.

It is very threatening to the way business is being done in the food industry and medical sciences because it exposes a reckless disregard for humanity in their ways of thinking, their ways of doing business, and the impact their products have on people, animals, and planet. It exposes their ugly intentions.

This book reveals many of the lies and deception that plague our print and airwaves. And it gives you big heaping doses of reality. In fact, I wouldn't be surprised if The Powers That Be felt so threatened by the information revealed in this story that they felt compelled to shut me up somehow.

As of 2018, only the biggest ideas covered in the book, like the microbiome's role in health or sickness, have become common knowledge. The majority of the ideas presented here are still not widely known or generally accepted (although that's changing fast). Most people still believe the propaganda.

Real world experience vs. scientific studies and peer-reviewed papers: Who to trust?

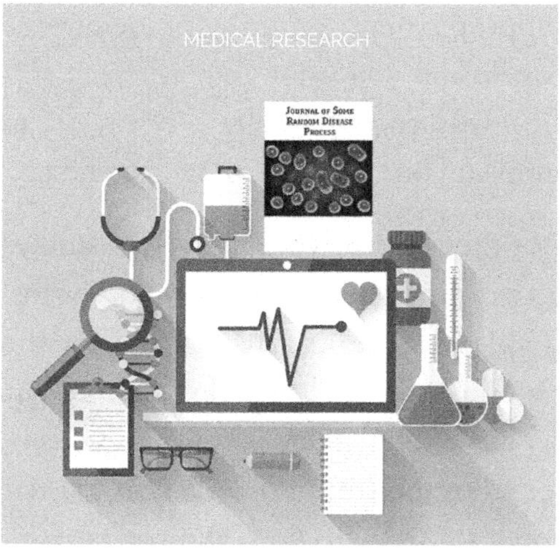

When you're at the forefront of human understanding, you have to take the cutting edge thought of others, make connections, hypothesize, and you have to draw conclusions.

You have to learn what has worked and what hasn't worked for others in practice. You have to remain open to alternative explanations. You have to be willing to speculate and be proven wrong in order to make rapid progress.

In contrast, although scientific studies and publishing are a standard part of the scientific method, intelligent people know scientific literature is only as valid and trustworthy as the intent of the people who backed the study, or are paying the bills of the researchers.

Meaning, there is good material you can rely on, and, just as frequently, there's a lot of dishonest literature meant to mislead, confuse, and distract. Just because a paper is coming from a seemingly credible source doesn't mean a thing. You have to look at the intent of the parties behind the scenes paying the bills.

To reiterate because this is such an important point, scientifically-based clinical trials and peer-reviewed papers are only as honest and reliable as the intention of the parties that funded the research. Smart people know this.

Anyone who's been around the block once or twice generally accepts the fact that most formal research is biased from "conception all the way to dogma" to favor conclusions that support some group's predetermined agenda. So whatever conclusion you want your research to show, if your wallet and will are great enough, there's a way to make the research appear to support your pre-determined agenda.

In other words, you can setup studies, and perpetuate ideas, that prove just about anything you want to, because it's become a simple, straightforward process for them manipulate study criteria, experiments, and conclusions, to make it seem like the data supports any pre-conceived conclusion you want, no matter how far from the truth it is. Deceit and trickery in science is easy when you know how to do it and you've got deep pockets.

Dr. Natasha:

> *"I diagnose by a clinical picture. I believe in clinical experience and clinical picture, much more than I believe in science. Our science has lost its way. More than 90% of scientific studies are commercially funded. They are deceptive. And the majority of us have no tools, and no training, to really decipher whether the study was designed properly, conducted properly, and then analyzed properly. It's very difficult.*
>
> *So there's a huge amount of information that leads you away from the truth. So I work with real people on a daily basis. I see what*

works and what doesn't work in real life. And if I need some scientific backing for any of those truths that I've discovered, I will find it. And usually those scientific studies are kosher... are correct. I believe in Mother Nature. I believe in real life, and real people."

Gut-Brain Secrets **is not a miracle cure, or the latest-and-greatest diet trend. Instead, it is the foundational elements of human health, as it applies to the gut and the brain**

I don't believe there is a one-size-fits-all diet, drug, or treatment protocol to reversing gut-brain problems... Not anymore. And certainly not in the years to come. As the healing community is becoming more keenly aware every day, everyone's body, exposures, and life circumstances are different in so many ways.

To complicate our healing efforts, the health threats we face every day are expanding from periodic sprinkles into non-stop showers. Meanwhile, our bodies' defenses and coping mechanisms are becoming exhausted in more and more people with each passing quarter.

As a result, no matter what positive changes you make to your diet or lifestyle, it seems there's always another threat around the corner, waiting to take its place – another environmental exposure ready to steal balance and function from whatever reserves you may have left. In other words, we're all affected. No one can escape the consequences. That is, unless you know the rules of the game.

So I advocate learning the basics of good health first, then taking steps to get a handle on these areas, if not get each area "handled." This ensemble of systems within systems includes: Increasing hydration in the cells, supporting the barriers of the body, evicting toxins from their hiding places, minimizing new toxin exposure, restoring communication networks of the body (e.g. neurotransmitters, redox signaling, and hormones), enhancing nutrient levels, expanding your energy supply, and, of course, bringing the microbiome into balance.

Fundamentals of Human Health (or Disease)

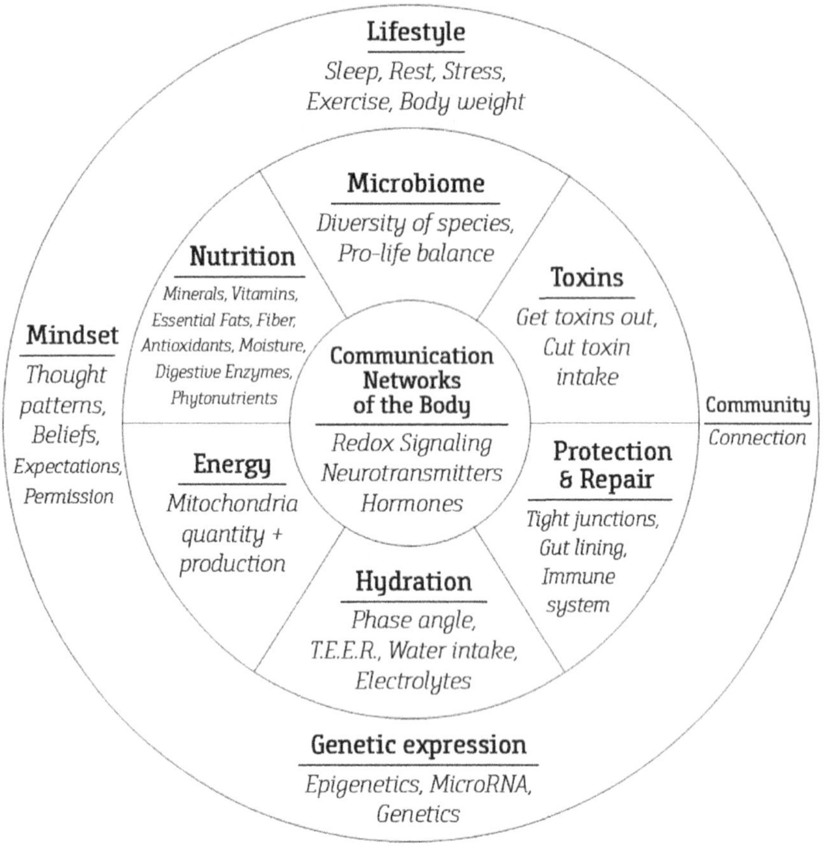

When you get a handle on these fundamentals of human health, disease is simply not needed or welcome in the body. Dysfunction is just totally out-of-place in a body that's clean, in-tune, and well-supported. In other words, fix the real source of your problems, and the symptoms that bother you cease to exist… instead of chasing symptom after symptom with "band-aids" and short-term solutions.

Simply put, *Gut-Brain Secrets* is your primer to learn these fundamentals of immaculate gut-brain health. It's your guidebook to help you get to wellville, and to keep you there. Strive not for perfection, though. You'll never make it. The more sensible approach is just to seek to *be, do,* and *have* better than you were the day before. *Gut-Brain Secrets* gets you started in the right direction.

We'd all be better off if this book were shorter and filled with good news

Unfortunately for us all, there's a lot of story to tell about the factors that create toxicity, imbalance, and dysfunction in the body. We're going to cover a lot of ground because the health hazards that surround us demand it. That's what the world's come to. And you better get used to it, because it doesn't look like disease rates are going down anytime soon.

In other words, this material is hitting closer and closer to home for more and more people. So you have to be prepared to fight for your own health, 'cause the deck is becoming ever-more stacked against the health of us all. Just ask any long-time practitioner. They'll say the symptoms and diseases they're treating on a daily basis are vastly more complex and stubborn than they were just a few decades ago.

Fortunately, there's plenty of juicy secrets to tell you straight from the cutting edge about why the body needs nourishment, how it adapts to problems, and what you can do to recover from diseases generally thought to be incurable. Knowledge and solutions are growing as fast as the threats are.

However, truth travels slowly when it prevents money from landing in deep pockets. So don't expect to get the truth, the whole truth, and nothing but the truth from mainstream sources. Frankly, for them, getting you hooked on what they offer is usually higher on their agenda than genuinely helping you.

At its deepest level, the real source of our health and environmental problems is the politics, power, and profits of a small group of people at the top

As you delve into the material, you may ask why so much ink needs to be devoted to covering politics, corruption, control, reckless profiteering, and the people behind the scenes pulling the strings. Can't we just focus on nutrition, biology and psychology, and forget the controversial stuff? Why devote so much space to information it seems we have little control over?

The reason is simple: This book is about the underlying causes of gut-brain problems. And you can't fully comprehend the true source of your problems, and how to go about fixing them, unless/until you understand who designed them, and why they set things up the way that they are. That's where 99.9% of efforts fall short today in some respect.

Until you understand (1) who is behind the system, (2) how they create dysfunction, disease, and destruction, and (3) why they do what they do, you're basically still trying to solve symptom after symptom, while the root cause remains (i.e. the perpetrator remains on the loose).

No matter how effective your nutritional/lifestyle changes are in reversing your GAPS condition, if you don't fix the real problem, you'll be spinning your wheels to some extent. You'll be putting out fire after fire in every aspect of your life, while the arsonist remains at-large, wreaking havoc wherever he sees opportunity to exert more control over your health, your life, and your pocketbook.

In doing that, new, more menacing, hazards and "solutions" will continue to pop up and expand as long as the real source is allowed to persist – as we're seeing today. As plainly as I can put this, elitist influencers, and their motivations, are the real source of our health and environmental problems.

Until you understand that, and we all put a stop to it, our problems will not only continue; they'll multiple and intensify at an accelerating rate. Just three manifestations we can put numbers to: Within one generation, cancer and autism are expected to affect half the population, while our fertility has already been cut almost in half.

Our once sturdy and supportive guts are now hurting us as much as they're helping us

Around 2010-2015, I feel we passed the tipping point where the average person could ignore this material and have some hope of getting by long-term without noticing any ill effects. Unfortunately, none of us have that choice anymore. The sources and severity of damage are expanding too rapidly to expect, or even hope, to be one of the lucky few to escape side effects for long.

You simply must understand how gut health works, or fails to work, along with its downstream effects, if you're to get healthy and stay healthy mid-to-long-term. It's no longer an option because the damage is already upon us. We've just failed to recognize the ways Our Corrupted Guts are draining the health right out of us and turning on states of disease.

Anyone who's up to speed on toxicity, the microbiome, and mental health can now say with 99% certainty you are currently being affected by a Corrupted Gut. You just don't know how its causes and effects are connected at this particular moment.

You don't know which symptoms you experience are caused by environmental factors that people have been blaming on "genetics"

… and what you can do about it (because there are solid solutions, with more in the works).

I say it like it is (and won't be suppressing any opinions)

I've made every effort to characterize my level of confidence in what I say fairly and accurately. If I'm not sure in places, I express reason for pause. When I am sure, I tell you in no uncertain terms. I fully expect time will reveal a small percentage of the ideas and conclusions to be off base, incomplete, or misrepresented in some way (if you disagree, present me with a rational explanation to the contrary, and maybe I'll change my mind).

My sources and I are bound to have made a few errors in judgment. But there's no doubt the vast majority of main points discussed in this book are right on the money. Feel free to doubt or disagree with my ideas or rationale. It doesn't matter all that much, because ultimately the science backing the gut-brain phenomenon is not opinion. We are seeing it rapidly become a self-evident matter of fact.

Indeed, the general consensus among leading, open-minded healers is already moving away from traditional medical and nutritional dogma, and instead supporting the pillars upon which GAPS, food-based medicine, and natural healing are founded. As of 2018, a number of basic GAPS concepts have hit the mass consciousness, like the impact that probiotics, antibiotics, and gut health have on the mind and body, as well as the harmful effects of toxins and sugar.

And a small but growing number of cutting edge healers who are leading the way are demonstrating clinically that many serious mental health disorders and physical diseases once-thought to be incurable are far from it. We can see a tipping point approaching on the horizon.

On the other hand, mainstream medicine is rapidly losing its effectiveness, and frankly its relevancy, in treating chronic diseases

Just look at the preceding "Fundamentals of Human Health" graphic again, and ask yourself how many foundational elements of human health regular doctors pay any attention to in evaluating and treating people – let alone have proficiency at treating. I count one area out of 11 that they routinely consider to be important: Lifestyle – which includes smoking, drinking, stress, sleep, exercise, and body weight. The other ten fundamentals they ignore almost entirely!

Progressive practitioners are not waiting for science to catch up to their knowledge and techniques. They're healing people now

Precious time is ticking by for countless children whose condition can be helped before damage becomes permanent. Time's a wasting for hundreds of millions of people whose lives are being devastated by ADD, autism, depression, anxiety, learning disabilities, and other mental conditions that are bogging down our health, our happiness, and our productivity.

Mental health disorders are tearing families and futures apart. It's time for action now, and we are making progress. Join us in learning how nutrition and gut health impacts our brain health and our quality of life. Welcome to our explosion of awareness, and our revolution of healing.

CHAPTER 1:

We're in a Matrix of a Mess

1

You've been lied to.
You're being manipulated. We've all been tricked

Unfortunately for us, most of the concepts our health and nutrition institutions have been selling us for decades is absolutely not true … not even close. Much of the information you've learned about health and how the body works is wrong.

Almost everything "science" and mainstream medicine teaches you about how to get healthy, stay healthy, and reverse disease is a lie. It's just not true. And in many cases, the efforts to misinform/deceive us have been carefully planned, well-financed, very intentional, and, unfortunately, remarkably effective.

In a word, you've been **BRAINWASHED** into believing what a handful of master manipulators want you to believe, because their machinations hurt you in subtle yet profitable ways.

You see, they don't want you thinking for yourself. They don't want you questioning their science, rationales, conclusions, doctrine, or dogma. They don't want you knowing the truth. And they don't want you controlling your own future and happiness (because then you wouldn't have any need for them).

On the contrary, what they've done is gain control over your health, your pocket book, and your family's well-being by telling you half-truths and pure fiction in a highly-organized effort to extract what they can from you, and make you dependent on "them."

And billions of people around the world are now paying the price in the form of outrageous healthcare costs, massive increases in serious disease, health annoyances affecting nearly everyone, high food costs (for quality food), oppressive wellness regimes, and most important for this exposé, runaway Attention Deficit Disorder and related psychological conditions.

Chronic disease has become America's biggest business

The drug companies, and the bodies they exert influence over, don't like it when people know the real reason disease happens. They hate it when people discover the root causes of serious disease, develop treatments for it, and try to tell others about it (at least when you succeed at doing it). They hate people knowing the truth, because it empowers you. It robs them of control. And it threatens their income.

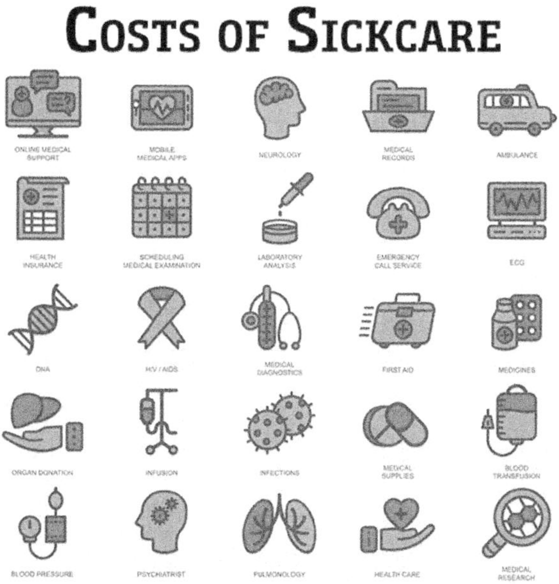

Fact is, the drug companies don't want you to die, and they don't want you to be completely well either. Both are bad for business. Rather, they like to keep you in a state of partial health – somewhere in between full health and death.

So they will do everything in their power to: ignore, deny, distract, obstruct, ridicule, regulate, legislate, suppress, buy-out, bury, kill, or otherwise defeat cures and treatments for illnesses that are too effective. They love partial solutions that keep you coming back for more. That is to say, **drugs don't cure. They just change the form and location of imbalance and disease in the body.**

Mainstream medicine and agricultural science leads us astray

Have you ever noticed the rise in ADD closely matches that of autism, along with many other psychological disorders? With several decades of "science" influencing our medicine, food industry, educational system, and media, we should have seen the rates of those conditions decline, or at least stay the same. But the exact opposite has occurred – they've skyrocketed. And there is a reason.

Hint: It's not a lack of willpower, modern conveniences, or a more sedentary lifestyle. Human physiology is still the same as it's always been. Genetics are still the same. Human nature is still the same.

The problems we're experiencing have more to do with what we're being taught, the foods we eat, the toxins we're being exposed to, the medical treatments we're getting, and the lifestyle choices we're unwittingly damaging ourselves with.

Whether through omission or commission, coercion or subversion, we have been misguided. Whether through ignorance or deceit we have been led astray by the very organizations we've entrusted to teach us and take care of our health, nutrition, and well-being.

But that is what it is. "They've" done what they've done. And they're going to continue doing what they're doing until they're stopped. And try as they might to convince us their motivations are the same as ours, they are not.

Their motivations are aligned with ours only to the extent it affects their image and profits – meaning government institutions are most interested in maintain their dominion over the people. And corporations are primarily interested in preserving their power and profits. Your health, and that of your family, is certainly one of their considerations. It's just not that high on their list of priorities.

But now we have to take our power back. We have to realize there are reasons mental health disorders like ADD, autism, anxiety, depression, OCD, and bipolar happen in the body. There are reasons physiological problems like heart disease, diabetes, cancer, multiple sclerosis, and food allergies/intolerances happen.

It's not bad luck in the genetic lottery. And it's not a mystery. We have to seek real answers, real truth, and real solutions. We have to take responsibility for our own beliefs, our own health, and that of our family. You, as the one person in the whole world who is most concerned about your own health, has free will to choose what to believe and what to do.

Only you can take responsibility for you own choices, your own health, and your family's future. Don't give that power to anyone. It's your right and obligation if you want to live your life to the fullest … if you want to live your best life.

Ken Rohla on how "The System" is rigged

> *"The medical system is really good at dealing with acute injury, but they're horrible at dealing with degenerative injury because their industry has been manipulated by the drug industry, the medical technology industry, and others…*
>
> *So I saw all these very well-intentioned people who were completely ignorant of natural methods of healing – nutrition and things like that. It was by design… for example, nutrition is kept out of medical doctors' training. Herbology and natural methods have historically been kept out of the medical system since its founding because those methods are viewed as not as profitable as conventional medical techniques using drugs and surgery.*

So there's this lack of awareness among well-meaning people who are perpetuating a system that's not sustainable and doesn't work well for degenerative illness. And yet they all believe that they're doing good for the population because of their programming and their beliefs… and so that's kind of a metaphor for what's going on on this planet.

You've got people at the upper echelons of power throughout history who have a better understanding of the nature of our reality… of matter, energy, science, and physics. They've intentionally kept the masses misdirected through systems they've put in place, that have evolved over time, and become more sophisticated, to keep people's awareness limited in a paradigm that benefits them, they can profit off of, that they can feed off of, and that they can control."

Two different approaches to nutrition and health: *Man's Way* vs. *Nature's Way*

Comparing two fundamentally different approaches to nutrition and disease can revolutionize your understanding of human health: One is based on ideas and approaches that support big business and political interests – backed by money, deception, control, and manipulation. The other: ground-level, real-world experience that supports life and Nature.

Most important to our discussion, the first paradigm is all about disease management and "what's in it" for a select group of people at the top, while the second paradigm focuses on balance, self-healing, energy, and sustainability the way Nature intended.

First, let's invent a fictional world where people in power like to control and manipulate the masses for fun and profit. In this fantasy land, anything goes so long as the people don't have knowledge, don't take responsibility, and don't take action. You can get away with anything so long as the people don't have a clue.

Approach #1: Man's Way

In the paradigm where a small group of ruling elite (call them "World Management") knows what's best for people and planet, large entities use a variety of tools and tricks to get their way. They might use "scientific research," publishing, marketing, PR, status, and authority to create the perception of reality. He who writes the checks makes the rules.

For instance, let's say a number of parties, like agriculture companies, had an interest in showing that a crop amendment, food additive, or genetically modified organism is safe. Or let's say they wanted to show that a food or nutritional component that comes from Nature, and that man has eaten for thousands of years, all of a sudden is bad for you (like saturated animal fat).

(Same schemes would work just as easily for drugs, surgical treatments, medical interventions, and even our very belief systems about the human body.)

Hypothetically, if these parties that stood to gain profits or power using less than ethical and honest means … How would they go about convincing the public that their idea is safe and effective – even if it's not?

Well, what they could do to start off with is find some research scientists who have impressive-sounding credentials, and/or who work with impressive sounding academic centers or institutions, and ask them if they want to accept a large block of funding to study a hypothesis.

'Here's 5, 10, or 25 million dollars to go study X, Y, or Z idea'. This grant will help pay for facilities, support staff, other programs, and personal lifestyle. Remember, scientists have bills to pay and families to feed just like everybody else.

And let's not forget the reputation and career path of researchers and scientists – their whole professional world to a large degree is measured by the papers they publish and the studies they take part in. Their worth and legacy as a research scientist is closely tied to the recognition they get, and who/what they're associated with.

So the special-interest parties find their people, and they offer them big grants to design a study, conduct a study, analyze the study, and hopefully publish the study. All the while the scientists – presumably they're smart enough to figure things out – they understand implicitly or explicitly which results/conclusions would benefit the special-interest parties that granted them the funding, and which ones would not.

How to brainwash the masses, Step 1:

Most important, these hand-picked scientists are "flexible" enough to figure out how to: (1) artfully craft the study criteria, or (2) the execution of the study, or (3) the data, or (4) the analysis of that data, or (5) the summary description of the study, or (6) some combination thereof to support whatever findings that would favor their "sugar daddies."

Generally speaking, if you earn money and spend it in this world today, you kind of know where your income is coming from – who pays your bills. So with some money to drive the project forward, the study pushes ahead.

Next what might happen in this world of make-believe is that follow-up studies, articles, PR pieces, commercial marketing messages, and academic curriculums (the books students are learning from) could all reference the results and conclusions of the first study, and support/re-affirm its findings.

Whereas all studies that failed to support the initial study's findings could be re-analyzed, rewritten, redone, buried in obscure publications, or simply never published. That might be a temptation if you stood to gain or lose billions of dollars based on what the study finds, and what the masses think.

These special-interest groups, through repetition over time, could build what appeared to be a consensus of opinion on the subject with industry collaborators (conspirators). With millions or billions of dollars at stake, lots of power/control to gain, and untold means and opportunity to exert influence over the prevailing opinion … they could all chip in to drown out the voices of skeptics (nonconformists). They could silence dissenters through multiple channels, using well-defined processes.

Examples of MYTHS founded and perpetuated on lies and deception:

- The (USDA) Food Pyramid.
- Cholesterol and saturated (animal) fats cause heart disease and stroke.
- "Everything in moderation". The idea sounds reasonable. Unfortunately, the belief that small amounts are harmless leads you away from accurate, independent thinking… and toward a way of thinking where you're more easily deceived and mind-controlled. Mercury, lead, glyphosate, and vaccines early in life are perfect examples: No amount of them in the body is safe.

How to build apparent consensus for bad ideas, the wicked way

Then for even greater impact, they could have those special-interest parties (e.g. food companies, chemical/agriculture companies, medical associations, government health institutions, prominent universities, and awards/recognition-giving groups) with their billions in airtime and unquestioned credibility, "amp-up" the fear and hopelessness by repeating messages of dire health consequences should people fail to do what they say.

Their giant commercial and PR engines could traumatize the collective psyche of the people by broadcasting messages of impending doom should people not reduce their "X", or buy more "Y" – no matter how wrong or damaging to people's health that idea is.

Our educational system could be deployed to ingratiate our leaders in all fields with those ideas – into that one way of thinking. They would repeat the message as gospel in textbooks, on government charts and guidelines, in news outlets, speaking opportunities, books, and lifestyle magazines.

Unwitting participants, just wanting to ride the bandwagon and make a buck, would undoubtedly get swept up in the confusion and find themselves joining in the chorus of voices singing the praises of a weight loss program that avoids the dreaded "X", or a diet high in the superfood filled with "Y". (Who can blame them? They're simply riding the wave of free publicity).

Government orgs, associations, and private foundations could regulate, investigate, stifle, and drive competing ideas out of the market. These parties could heighten the perceived credibility of their own ideas by bestowing their chief spokesmen and women with awards, collaborative associations, recognition, and lots of airtime.

So you think that could actually happen?

You think people could actually be so cruel and manipulative to their fellow man to lead them astray like that? And what would this imaginary world be like to live in? What would the end result be if these crazy situations were allowed to run their course, more or less unhindered?

Answer: We would be brainwashed into believing our institutions, out of the goodness of their hearts, were looking out for our best interests, not that of their benefactor/controllers (the ones paying their bills).

We would find ourselves in a complex web of influence and control that spends 99% of its research and marketing budget on a disease-centric model, instead of teaching people how the body really nourishes, protects, and heals itself with food and healthy lifestyle choices (thankfully, as the truth comes out, it's forcing us to rethink our paradigms).

And we would have a population that got so backwards in its thinking, so perverse in its lifestyle choices, that it actually thought Nature is stupid, fragile, and doesn't know what it's doing.

We would have even the most intelligent people believing Mother Nature made a whole bunch of mistakes in its millions/billions of years of development, and that man's science and industry can outsmart, out-engineer, subdue, or otherwise basically conquer Mother Nature. Hah!!! That is a laugh. When you know the truth, that is a real "laugh-'til-it-hurts" riot.

When you look at the complexity and performance of the human body … when you look at the engineering and adaptability of life forms on this planet … when you understand the cycles of life and death … you realize, at this point, **we as a society are getting more things wrong in our agriculture, medicine/healthcare, food supply, commerce, and energy production than we're getting right.**

We are mucking it up big time. And our shortcomings and missteps are beginning to become so desperate, we cannot ignore them any longer – pollution, environmental degradation, natural resource depletion, moral decay, runaway debt, rampant discontent, lack of responsibility, infertility, and, germane to this discussion, practically universal disease and dysfunction of brain and body.

We need to wake up. We are headed on a collision course with self-destruction, and we're virtually asleep at the wheel

But we're not going through this examination of mass mind control and problem identification just to blame any particular group. It's to show you how, given the right opportunity, certain people and groups might use their positions of power and control for their own benefit.

More precisely, "they" might not place your health, happiness, and well-being at the top of their priority list in their thoughts and actions. They might tend to spend more time worrying about themselves, and their own interests, than you and I, once they've tasted (i.e. gotten drunk off of) power, privilege, and control. I'm just saying this could happen from time to time.

But that is what it is. No need to dwell on it. What's important is, believe it or not, that long list of catastrophes we're now facing – they're all related to a Corrupted Gut in some way.

Interestingly consistent with what we just talked about, a Corrupted Gut causes man to act out more of his animal instincts by suppressing higher brain functions that would normally control his base impulses.

In other words, corruption of the gut is increasing the very conditions in the brain, and in our actions, that further worsen gut-brain problems. It's the perfect storm of man's lust for power and control meeting man's primitive urges from the hind brain. And it's wrecking the planet and all the people on it.

Approach #2: Nature's Way

One example how working with Nature, rather the against it, can help mankind discover (re-discover) practical, functional knowledge we can use

Weston A. Price, DDS was a dentist and prominent health researcher in the 1920's and 1930's. He travelled all over the world studying the relationship between nutrition, dental health, and physical health in indigenous peoples 'before' and 'after' a Western Diet had been introduced.

Among a plethora of fascinating discoveries he cataloged in his travels, native eskimos told Dr. Price that their sled dogs could not pull a sled all day long when they fed them regular fresh fish. On the other hand, their dogs happily pulled hard all day long when their owners fermented the fish first.

So the eskimos concluded, if fermenting increases the stamina of their dogs, then it must be enhancing the nutritional value of the food somehow. It must be providing positive benefit to the physiology of the dogs. Of course, they didn't know how or why fermenting worked. But it did work.

They didn't notice any unnatural side effects. It never failed to work. It seemed to align with nature. So they stuck with it. And that is the miracle of Nature in action – a way of thinking that The Powers That Be have been trying to ignore/suppress/deny for decades.

The eskimos didn't need to pay someone to study the science behind their observation. They didn't need to know the mechanisms of action in the body. They didn't look to any government institution to proclaim the process safe and grant formal approval.

They just made the assumption, 'if it's good for our dogs', it must be good for us too. They trusted that this wasn't an accident – Mother Nature actually knew what she was doing. And so they went with it.

Likewise, seeing what works and what doesn't in the real world is a much more expedient and direct route for the holistic community to take when they haven't got millions or billions to back their ideas, like the food industry does.

Much of the material to follow gives you example after example of Man's Way getting it dead wrong over the last several decades, and Nature having the truth and the way all along.

Myth: If something's good, I would have heard about it already

Nope… Being a marketer myself, I know the process of getting people and groups to believe ideas is very carefully planned, orchestrated, and executed. And, let me tell you, educating people and changing their minds can be a monumental effort. It's usually VERY expensive and time consuming.

It's getting easier with the Internet, social media, and the breaking down of traditional channels (i.e. media/publishing, science, education, medicine, banking, business, religion, law, and politics/government). But the techniques invented by Edward Bernays (the father of Public Relations and propaganda) in the 1920's and 30's to build credibility and craft/manipulate public opinion … they're still alive and well today. And they're still being used with shocking success by the people and organizations you've been taught to look up to.

Because Bernays' methods are built on innate social rules and human nature, you never see it coming. Remember, we may be human, but we're still part animal. You may think you're thinking for yourself, and making your own decisions. But the reality is, your thoughts and actions are largely determined by factors outside your conscious control.

Groups are led by emotion

Groups are controlled by a reactionary, one could say pack-like, mind rather than a mind that's individual, proactive, or self-determined. That's why it takes a monumental amount of organization, effort, and usually money to get a group to achieve something constructive or rational (like herding cats).

Yet people can be coaxed into supporting the most heinous acts of destruction, indecency, and absurdity by triggering their emotions. Spin doctors and PR people know this. Hitler knew it.

In a nutshell, most big ideas and accomplishment you see in the world are conceived, paid for, and propagated in order to promote some big organization's agenda. On the other hand, the individuals who want to give and serve others seldom have the strategies, the organization, the money, the backing, and, most important, the willpower to build a public consensus and effect change on a large scale.

The people that truly support freedom, empowerment, humanity, and service to others rarely have the techniques or the money to fight The Establishment. Their good nature leaves them vulnerable to attack, so they almost always get slaughtered when they go head to head with The Powers That Be.

It's very interesting: When you study social dynamics at the individual level, and at the group level, you realize **the bigger the lie, the easier it is fool people into believing it.** Why? Because individuals have some ability to think and reason for themselves; whereas, groups have none. Groups are led by "groupthink."

I'll say it several times: Medicine/healthcare is a business. It's big business. It's a business masquerading as science and as a profession. In fact, healthcare is the biggest industry in the world. Likewise, food is a business... And Big Food controls most of the nutrition education we receive today.

Bottom line: We know intellectually that the goals and actions of The Powers That Be are motivated by profit and control. But we want to believe so badly that the world is a fair place, and that our leaders care about humanity and the planet, that we'll gladly go along despite all evidence to the contrary.

Nothing inherently wrong with profit or control; it's the intention and methods that do the damage. What I'm saying is, most of the information you've been fed is either partially true, an intentional distraction from the real issue, or is a complete fabrication.

So now it's up to you to seek the knowledge and raise your responsibility, because no one is going to care more about your health and well-being than you are. Throughout this book, we'll examine many popular beliefs and paradigms that blow the lid off the lies and deception being thrown at the population from the top down.

Our saving grace: Every problem has a solution

The good news is, you don't have to get depressed, feel hopeless, or live in fear when you hear about all the problems we face with our personal health, as a society, and on a planetary scale.

That's because the way the universe is setup, each disease and problem must have an equal and opposite force counterbalancing it. It's a simple matter of polarity. And it's a rule of Nature – a foundational law – upon which our universe is built.

So for the negative to exist, there must be a positive. There is darkness, and there is light. There's hot and there's cold. For every yin, there is a yang. Highs can't exist without lows. And so on.

Likewise, the moment a problem is created, the universe (consciousness) simultaneously calls the solution into existence as the problem's polar opposite. Whether you're talking about diseases, systemic problems in society, evil plans, or suppressive people, the problem cannot exist without a solution. Of course, problems are usually crafted, executed, and publicized better than solutions are, because of the nature of who's usually behind each.

But as bad as things seem to be, you can be sure that a solution for each problem is out there for those who seek it. So you don't have to accept someone else's idea (construct) of what you're life and health are supposed to be like. You don't have to be a victim of someone else's reality. You can create your own.

But you will need to gain some knowledge. You will have to learn a few new tricks

Solving major healthy challenges will probably take some doing to apply in your life. But never forget, the force that created the problem in your life also created the solution.

Moreover, the universe doesn't force you to experience any more than you can handle. So when you're truly ready for answers, and you seek them out, you can count on finding the right teachers and solutions. Indeed, there's an old saying, "When the student is ready, the teacher will appear."

Staying in the right frame of mind on your journey is crucial to not getting caught up obsessing over the problem and accidentally attracting more of that into your life. Instead, learn about the 'who, why, and how' of the negative forces affecting your health.

Then release them of their emotional charge. Let go of the fear you may have for them – of their power over you – knowing that the solution is out there. Then, for best results, redirect that attention and positive energy toward the solution, and draw that health and prosperity into your life like a magnet.

Ken Rohla on the importance of mindset:

> *"Great masters throughout time have taught this concept of being unattached. And the reason is, because when you create your reality, yes, you're responsible for everything in it. And you can take that to a great degree with things for example like ho'oponopono, which is a Hawaiian kahuna concept of taking responsibility for everything in your life that you encounter, and asking forgiveness for it. And when you do that, everything around you changes with great effect.*
>
> *On the flip side of the coin, you've got this New Age concept of everything's just love, light, and angels. And, of course, the Law of Attraction says you should focus only on what you want to attract. Therefore, you shouldn't look at the dark side.*
>
> *Well, the problem with that is, you don't want to be dwelling on the dark side of things. But you need to be aware of it, so you can navigate through this reality. And if you truly 'get' at a deep, emotional level that you create your reality, and that you're empowered, and that you're living consciously, then the dark side can't affect you.*

You will always have a counter to whatever things the dark side is coming up with. And so it's very important to face your dark side, or whatever these negatives things are in your reality, and to understand them. And through understanding them, then you can do what you need to do to counter-act any possible effects they may have on you, while not dwelling on them.

You can be aware of them, you can understand the nature of them, and then you can focus on the positive. Focus on what you want to create, and you'll attract that. But ***living in denial of the dark side is every bit as disempowering as just focusing on the positive.*** *You've got to be aware of it, but not focus on it.*

When you create your reality, and you have the power, then these negative things don't have power over you. For example someone asked me on an interview yesterday, 'Don't all these negative things in the world make you depressed?' Absolutely not, because I know the solutions are out there. I've seen them. I've witnessed them.

So I'm not depressed at all. I think in the end, we're going to pull everything out just fine. I'm out there in the world sharing with people these solutions, so that we can all change it."

Knowing the truth about what causes health or disease is quickly becoming one of the most valuable lessons you can learn in life

It's one of your greatest assets in a life well-lived. Especially for ADD'ers, the truth about what causes mental dysfunction can change your life, because without physical health AND a well-functioning brain, you're going to get a fraction of what you can out of life. Without a brain that works right, your life is one big compromise.

One of the goals I had in creating this book is to give you a brand new way to think about how the body works, so you can reverse your psychological and physical health challenges and build better health for you and the people you know. Knowing what impairs brain function and what enhances it – those fundamental health concepts – and what you can do about it, starts right here.

CHAPTER 2:

A Mother's Determination to Heal Her Son Starts a Worldwide Awakening

2

When Dr. Natasha started talking about the gut-brain connection 20 years ago, pretty much everyone laughed at her

Not anymore. Her teachings of yesteryear that once seemed peculiar and speculative are quickly proving to be scientific fact. Nowadays, there's countless millions pouring into scientific research and the development of commercial products that support the health of our guts.

The gut-brain connection seems obvious now that esteemed institutions, doctors, and media are talking about it – while big money interests have hopped on the bandwagon with their probiotic supplements and cultured foods like yogurt and kombucha.

You see, all truth goes through three stages: First, it's ridiculed. Then, it's violently opposed. Third, it's generally accepted as being the obvious truth – like everyone knew it all along. As a result, Dr. Natasha suddenly (after twenty years) finds herself at the forefront of a worldwide renaissance that's revealing the all-important connection between the gut, the brain, and the body.

What inspired Dr. Natasha to get into medicine, and then gut-brain health?

At 18 months of age, Natasha came down with a very serious bout of food poisoning and almost died. She lost a lot of weight and had terrible digestive problems. So her grandparents took her to their village in the Ural Mountains where they had a small family farm with livestock, poultry, and a garden.

Her grandmother, who was a village healer, herbalist and therapist, took her in and cured Natasha. And she cured her with diet … with food. As she grew up, Natasha would spend school holidays in the village with her grandmother, observing how she would cook, how she prepared food, and how she used food as medicine.

Natasha would watch her grandmother help villagers with not only their health problems, but with their family and life problems as well. And she always had good advice. Natasha was very impressed seeing the world of good a healer can do in the lives of others. So she decided to follow in her grandmother's footsteps.

In 1984, Natasha Campbell-McBride graduated with honors as a medical doctor from Bashkir Medical University in Russia. She later earned a post-graduate degree in neurology, and practiced five years as a neurologist, three years as a neurosurgeon.

Just a few years into her career, she noticed what seemed to be a strange coincidence happening in her patients

She noticed that most, if not all, of the patients she saw for neurological conditions also had very severe digestive problems as well. But, much to her dismay, Dr. Natasha found mainstream medical specialists usually don't talk to each other in an effort to gain insight into how their colleague's systems work, and how they might correct problem conditions. Rather, specialists tend to treat the body's systems as a separate and unrelated "departments" of the body.

In her case, neurologists don't talk to gastroenterologists to make connections between the brain and the gut. As a result, Dr. Natasha's observation that psyche patients have some of the worst digestive problems – that might have remained coincidence and conjecture in her mind if life hadn't thrown her a curveball:

Three years after her first child was born, things went horribly wrong. His development, behavior, and social skills grew increasing abnormal. He was diagnosed severely autistic. And when she discovered that her own profession had nothing to offer them other than coping strategies, as a mother, she was not prepared to accept that.

So she threw herself into intensive study into the very nature of how autism originates in the body. She went back to university, studied everything she could study, read every book she could find on the subject, and spoke to hundreds of people.

Eventually she came up with answers and solutions to reverse autism in her child. And with his unwavering effort, she cured him. Today, her son is a healthy, happy young man – fully recovered and leading a normal life. And they've forgotten about autism long ago.

This trial by fire is what led Dr. Natasha on her personal and professional journey. She earned a second post-graduate degree in human nutrition from Sheffield University in the UK. As she was working with her son to correct his challenges, she returned to work and began treating other children with autism, and other learning disabilities, with great success. News spread, and people with all sorts of psychiatric condition began to seek her out.

Today, Dr. Natasha's many years of education and clinical experience have culminated in the creation of a program that explains and corrects the root causes of most modern psychological and physiological diseases. It's called G.A.P.S. GAPS stands for both Gut and Psychology Syndrome, and Gut and Physiology Syndrome.

GAPS is revolutionizing the world's understanding of chronic, degenerative disease, because it explains the actual mechanisms of action by which many mental, physical, and autoimmune challenges originate in the body.

GAPS reveals the sequence of events that happen in the digestive tract that cause a cascade of effects that manifest as modern disease. And the GAPS healing protocol tells you how to use food, probiotics, and detox protocols to heal the root causes of modern disease at the source.

Dr. Natasha is now recognized as one of the world's leading authorities in treating children and adults with learning disabilities and other mental disorders. As a pioneer in the field of the gut-brain connection, she is an in-demand speaker, author, and advocate for the natural treatment of autism, ADD/ADHD, dyslexia, dyspraxia, depression and schizophrenia.

She now lives in the UK. Her books include *Gut and Psychology Syndrome*, and *Put Your Heart in Your Mouth*, and *Vegetarianism Explained.* And she's working on a third book *Gut and Physiology Syndrome*, which is expected out in 2018.

GAPS was born out of Dr. Haas' Specific Carbohydrate Diet (SCD)

Renowned American pediatrician Dr. Sidney V. Haas invented the Specific Carbohydrate Diet in the first half of the 20th century. He treated over 600 celiac disease patients with excellent results: After at least a year on the regiment he saw no relapses, deaths, crises, pulmonary events, or stunting of growth in his patients.

He and his son, Dr. Merrill P. Haas, published the results of Dr. Sidney's research in 1951 in the medical textbook *The Management of Celiac Disease.* Shortly thereafter, The Specific Carbohydrate Diet, as described in the book, was generally accepted as a cure for celiac disease.

Unfortunately, in those days, celiac disease was not well understood or clearly defined. And many disorders of the gut were categorized as celiac disease. Later, it came to be defined as a gluten intolerance, which then excluded a large percentage of digestive disorders from the definition of celiac disease.

Problem was, because the medical community thought gluten intolerance and celiac disease were one and the same, the gluten-free diet was soon proclaimed to be the treatment of choice for celiac disease. So the SCD diet was dismissed as no longer being relevant. That effectively ended all research into finding treatments for inflammatory gut conditions that fell outside the box of celiac disease, because they thought they had the cure.

Following the mix up, the Specific Carbohydrate Diet would have been forgotten, were it not for a desperate mom that found success treating her daughter with the diet. Elaine Gottschall's little girl suffered from severe ulcerative colitis and neurological problems, and went to see Dr. Haas in 1958.

Two years on the diet turned out to be a smashing success. Elaine's daughter was completely symptom free, energetic, and thriving. Eternally grateful to Dr. Haas and the diet for giving her daughter the gift of excellent health, Gottschall took up the mission of helping others using the diet.

For years, she studied the biological effects of the diet on the human body, which culminated in her book *Breaking the Vicious Cycle: Intestinal Health Through Diet.* Since then, she and her book have helped thousands of people overcome Crohn's disease, ulcerative colitis, celiac disease, diverticulitis, and other digestive disorders.

Along the way, many children have also reversed their behavioral difficulties, including autism, hyperactivity, and night terrors. All of which helped lead Dr. Natasha to refine the SCD diet, over years of clinical practice, into her own version specifically tailored to reverse today's gut dysbiosis and mental health issues. But curiously, it was her patients that dubbed the food-based diet they were following the "GAPS diet."

CHAPTER 3:

The Wisdom of Nature vs. the Fallibility of Man

"Illnesses do not come upon us out of the blue.

They are developed from small daily sins against Nature.

When enough sins have accumulated, illnesses will suddenly appear."

— Hippocrates

(Based on the work of Barbara O'Neill, ND)

3

Pathogens serve a purpose just like probiotics do

Every type of microorganism on this planet is here for a reason. Every one of them has an important role to play in keeping the earth's ecosystem moving and prosperous. That includes microorganisms we traditionally think of as disease-causing, such as viruses, molds, yeasts, fungi, and bacteria.

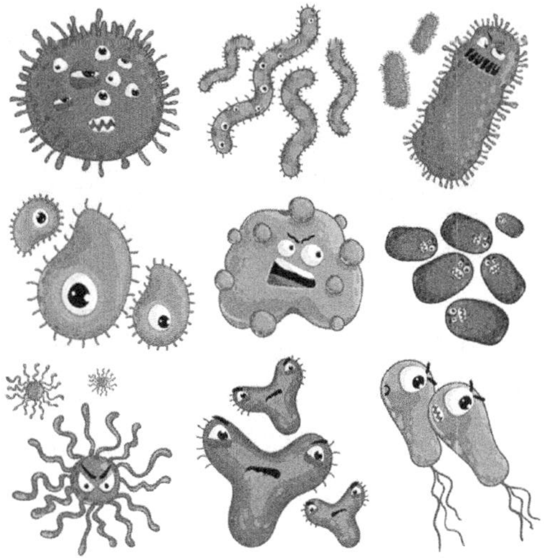

As discussed throughout, bacteria in particular play far more pro-life roles in the food chain than we've traditionally given them credit for. Bacteria control

- **Nutrient cycling in soil.** The growth and sustenance of naturally-grown plants is run by bacteria.
- **Short chain fatty acid production in pasture-grazing animals (e.g. cows).** Bacteria convert fibrous plant material into fat in the stomachs of pasture animals.
- **The digestion of carnivores that eat the herbivores.** Bacteria control the digestion of most creatures on earth.
- **Regulate the health of apex consumers.** Bacteria are essential to nutrient production and absorption, immune

system modulation, and gene expression of omnivores at the top of the food chain.

So it's not hard to see why beneficial bacteria build, maintain, and support virtually all life on planet earth

Bacteria literally regulate the health of the entire food chain from start to finish. *So when your bacteria ain't healthy, you ain't healthy.* Indeed, when the bacterial population of any ecosystem is skewed toward the pathogenic, or is nonexistent, that ecosystem suffers in many ways that are becoming abundantly clear to anyone who's paying attention.

But the concept people may not fully understand is what the other guys, the "disease-causing" pathogens, are here to do, and why life on earth can't exist for long without them. And that is, their purpose in life is to break down and decompose plants and animals after they die. Pathogens are the garbage collectors and recyclers of the planet.

They return the mineral and nutrient-content of dead plants, animals, and all forms of toxic material back to the earth (i.e. the soil and water) so they can be used again in the next "go-round." This 2nd half of the cycle of life is not-so obvious, because it seldom gets the same attention as the way in which nutrients make their way up the food chain.

One reason is because death and decay is not a sexy topic for our media to cover. It's not as appealing for our science and industry to study. Nor is it as fruitful for commercial advertising to reference. So nobody really talks about it much. Simply put, death isn't pretty, and decay doesn't sell. So neither get the same press that brighter topics do.

But death and decay do create the raw material for the next cycle, so it's crucial to understand why pathogens show up in the first place. This is a very important point:

Pathogens are called to duty when *disease conditions* are present in the environment

In this case, pathogens thrive when disease conditions dwell in the body. That's when disease is invited to show up. That's when disease sticks around and makes its presence known.

What medical science neglects to tell us is that barring man's intervention, pathogenic organisms are always present in small numbers on every root structure, every leaf, every piece of fruit, and every scoop of soil. All the animals that inhabit the earth (including us) host pathogens all over the inside and outside of their bodies. They're in the rivers and seas. They're in the air. They're everywhere.

In a healthful environment, pro-life organisms predominate, and pathogens don't cause any harm. You would never even know they're there. But there are two situations where pathogens multiply and grow abundant: One, when disease-promoting conditions in the environment tell the microbiota it's time to recycle dead plant and animal matter. And, two, when the probiotic flora that normally keep the pathogens in check are suppressed in some way.

They usually go hand-in-hand, because when an animal dies its probiotic flora deteriorate along with it. Likewise, when your gut flora is weakened you become nutrient deficient and disease processes kick in. In other words, the present state of the average person's internal environment increasingly looks like it's dying and in need of decomposition to their microscopic housekeepers.

So what causes these disease conditions that trigger the overgrowth of yeasts, molds, fungi, viruses, and bacteria?

That's the golden question that has far-reaching implications. Here are a few conditions that precede and contribute to disease:

- Poor oxygenation of the blood and tissues.
- Overly acidic conditions (the first two usually go together).

- Buildup of positive charge (i.e. too much oxidation and not enough antioxidants).
- Buildup of toxic metabolic waste products.
- Nutrient deficiencies (which result in the disruption of biochemical processes).
- Buildup of environmental toxins, such as heavy metals, industrial chemicals, crop amendments, hormone mimics, and pollution.
- Chronic inflammation.
- And lack of diversity in the microbiome.

The common theme with these conditions is that they're either a (1) digestion/metabolism problem, an (2) elimination/build-up problem, or (3) an imbalance issue. All of them are the foundation upon which disease comes to fruition. They are the ingredients, or recipe, for disease to materialize.

Conversely, when you're getting the nutrients you need, eliminating waste properly, and your biochemical processes are well balanced, you're naturally free of disease. A good example is cancer. Cancer can't survive in the presence of plenty of oxygen, the right nutrition, a clean environment, and cells that are replicating properly.

To reiterate, pathogens proliferate when unhealthful conditions invite them to be there. They show up because these conditions in nature indicate life force is low and it may be time to decay the plant/animal back to the earth. Pathogens grow out of control when they find an environment that supports these unhealthful conditions.

The other condition that's slowly, but surely, increasing pathogen dominance

Pathogens will proliferate when friendly flora isn't there to protect you, like after taking antibiotics, suffering a major illness, or a major infection like cholera. The reason is, probiotic life is naturally superior because the universe favors life.

On a "level playing field" probiotic life forms are stronger than pathogens and can out-compete them with more powerful weaponry (anti-microbial substances). However man's interventions have been upsetting this delicate balance of Nature for decades now.

And that's why you usually don't see pathogens grow out of control in your body as a result of "infection," as medical science believes. Instead, they proliferate because your good guy bacteria are weak, your immune system is compromised, you're malnourished, your pH is off, and/or your internal environment is somehow upset.

So you see, pathogens are not the bad guy. Even though we've described them as being disease-causing, they are not evil. They're simply the other side of the cycle of life that we don't like to talk about, write about, research, or educate our population about.

Disease-causing microorganisms are Nature's way of culling weak and infirm cells and organisms, and removing them from the food chain, so their resources can be put back into the ecosystem. Pathogens have a purpose that we need and benefit from.

They even do some good things in the body. Many of them have some special talent only they can do – like synthesizing certain vitamins, producing some antibiotic substance to suppress other pathogen populations, or detoxifying select chemicals/heavy metals/toxins.

In other words, Nature knew what it was doing when it designed probiotics, pathogens, and the delicate balance between the two – in our bodies, in the soil, and in every living creature.

As for their role in your own health, when pathogens show up in your life prematurely, you can think of them as early warning signs to change your diet, change your lifestyle, heal your gut, and change your ways – before you don't have a choice in the matter and they seize control of the situation.

Microbes transform themselves according to what's needed in their environment

Royal Rife's work confirms this, and Dr. Robert O. Young concurs (author of *Sick and Tired?*). They showed that microbes have the ability to transform themselves from bacteria to fungi to virus, and back again, in response to their environment (called pleomorphism).

When an organism is healthy, microbes will take one form, and when the organism dies and is in decay, the microbes' role will change from one that supports life to one that helps break down the dead plant/animal matter to be recycled back to the earth.

In other words, microorganisms sense their surrounding environment and adapt themselves to a structural form most suitable to live in synergy with the prevailing conditions. This is Nature's way of facilitating the cycle of life and death.

However, this metamorphosis does not happen instantaneously. And it's not readily discernable to the casual observer, thus mainstream medical science has very little to say about pleomorphism. As a concept, the whole idea of pleomorphism conflicts with their agendas, so anything they know about it they're keeping to themselves.

Germs aren't the enemy. Poor environmental conditions and imbalances are

Today, no thanks to those pushing sanitizing gels, antibiotics, and sanitizing practices we live in a "germaphobic" culture. But as this publication suggests, germs are not the primary source of disease.

They're merely a pawn in the process. Nature designed them to play a role. So they're found at the "scene of the crime." However, they're not usually the bad guys.

When your inner ecosystem is weak, imbalanced, malnourished, toxic, or otherwise vulnerable that's when germs can get in, establish themselves, and have a field day. When your system is strong and in balance, germs can't grow and proliferate. More than anything, this means having an abundance of beneficial bacteria colonizing your intestinal tract.

Many hospitals today are feeling the pain of opposing Nature in this way, as more and more of them are becoming breeding grounds for drug-resistant germs

They try mightily to completely sterilize every surface in the building, biological or not, over and over again, with their chemical disinfectants. However, what happens is, pathogens are the first to grow back after each cleansing.

So without good guys there to protect the territory against pathogens, all these institutions succeed in doing is creating unguarded terrain for the nastiest microorganisms to set up shop and breed as best they can in the relatively infertile grounds of hospital furniture, fixtures, instruments, office supplies, and workers' clothing.

The medical system seems to be slow to understand that killing everything carries substantial risks, and is not necessarily the answer they're looking for. They think more frequent use of hand gel, and more thorough sterilization, are the answers to eradicating drug-resistant organisms in their facilities.

But it's flawed thinking. It's an incomplete, and increasingly problematic, paradigm. As a result, some hospitals have even been forced to shut down, demolish, and rebuild their facilities because all attempts to eradicate the multiple-drug-resistant pathogens have failed. All attempts to get infection rates down were unsuccessful so they basically have to hit the reset button and start all over again.

Old world infectious diseases and modern syndromes may, in fact, have a common cause

Many diseases that ran rampant in the old-world appear to be an overgrowth of infectious microorganisms you can see under a microscope. They're caused by microbial infections that thrive(d) under horribly unsanitary conditions, as we'll discuss later on.

On the other hand, most modern diseases are a collection of symptoms we group into a diagnostic box and label as a disease. In other words, **modern diseases are a bunch of effects and outcomes we call symptoms, not the sources and causes. Some of them are associated with microbial overgrowth, but most are not.**

Point being, something upstream from both old and new diseases is causing its symptoms to occur, and often a pathogenic organism does not factor into the equation. Instead, it's unhealthful conditions that allow the disease to take place. Diabetes and heart disease are two examples.

So pathogen growth is a side effect, or a consequence, of those conditions, not the driving force that creates the disease. And for that reason a diagnosis these days tells you little to nothing about what's causing the situation – at least as far as regular doctors are concerned.

Louis Pasteur's "germ theory" vs. Florence Nightingale's "wretched environment" disease model

Louis Pasteur made the germ theory famous in the 1850's. He studied the way tiny organisms grow and behave under a microscope. And he concluded germs cause disease. But if germs really cause disease then surely all we have to do is find something to kill germs and we could eliminate disease. Right?

Well, not exactly. We have plenty of things to kill germs, yet disease is skyrocketing with no end in sight. Indeed, the more we try to wipe out microorganisms through antibiotic practices, the more rapidly pathogens become resistant to drugs and spread.

Florence Nightingale's "wretched environment" model (my description)

As Pasteur was conducting his research, Florence Nightingale, the founder of modern nursing who was born to very wealthy parents, was sent to look after soldiers injured in the Crimean War in the mid-1850's.

When she arrived, conditions in the hospital were appalling. The plumbing didn't work. Raw sewage polluted the corridors. The kitchen was filled with rotting food. Soldiers were waiting days or weeks to be attended to. And the death rate was 50%. Soldiers had a better chance of surviving the battlefield than that hospital.

She had her work cut out for her and volunteered to help in any way she could. She offered to clean up the filthy facilities she believed to be causing the terrible mortality rates. But the doctors refused to let her, and her group of 38 nurses, clean up the wards.

Instead, they sent her to the kitchen. So nurse Nightingale did what she could. She scrubbed the kitchen. She telegrammed her father and had him send a shipload of clean linens, clean clothes, clean mattresses, clean bandages, and a cook.

Two and a half weeks later, yet another ship of wounded came in that pushed the hospital further past its already stretched limits. The staff was already working at capacity. So the doctors reluctantly agreed to let Florence and her nurses into the main hospital.

The first thing Florence did is get her nurses to scrub the wards down and open the windows to let fresh air in. She made all the doctors wash their hands between operations. She got the plumbing fixed. She put the clean linens into use. She cleaned up every nook and cranny in the hospital where germs can live, and within six months the mortality rate had dropped to 2%.

When she went back to England, Nightingale was hailed as a heroine. But she objected. She said all she did was improve the hygiene, the sanitation, and the nutrition. Florence Nightingale knew that germs don't cause disease. Germs are the result of unhealthful conditions.

Florence Nightingale knew what our medical science is just now beginning to accept as a reality: Pathogens can't thrive in a healthy environment. Whether through cleanliness, a strong immune system, or the presence of beneficial probiotics, germs have no purpose or place in healthful conditions. Pathogens struggle to live in a healthy body. And that dramatically reduces the "foothold" by which pathogens can gain traction in the body and become disease.

Ants got no respect

A real life example from my own experience: I live in a place surrounded by trees and shrubs. It's basically a forest. And no matter what anyone in the community does to try and keep them out, there's always going to be a lot of bugs and forest critters amongst us.

Naturally, a few of them are bound to find a way to get inside and pilfer anything they can eat. So no surprise, the summer after moved in, I found some tiny ants marching around like they owned the place.

Now having learned what I learned from Barbara O'Neill, I knew ant poison and ant traps are one way to try and get rid of the unwanted guests. That's the first method that comes to mind when most people think about pest control.

But a far more practical and effective way is to thoroughly clean the place and remove all sources of food and filth that attracted the ants in the first place. I did it. And you know what, it worked. I've seen scouts from time to time on what seem to be reconnaissance missions. But they don't stay.

So basically, instead of attacking the ants, which are the result (the effect), I cleaned up the food and dirt, which was the real problem (the source). I removed the reason they were there in the first place. And once the things that attracted them had been removed, they had no reason to be there. So they left and sought sustenance elsewhere (in someone else's home). In short, it worked.

Thus we come to this book's "Big Idea"

The moral of the story, which we're going to approach from many different angles, is that **symptoms of most modern diseases have no reason to be in the human body when the conditions are not right for them to exist.**

In other words, symptoms – such as food sensitivities, heart disease, arthritis, and attention deficit – are not the real problem. They are not the source or cause of the problem. They are the body's attempt to try and tell you something (with the only tools it has at its disposal).

They are a <u>re</u>action. What we call "symptoms" are really the product of some trouble or malfunction the body is having elsewhere. And when you fix the real source of the problem, the symptoms vanish. This will be a recurring theme throughout…

Nature favors life

And life proliferates to the fullest extent that it can, wherever that may be – in soil, in the human body, in the atmosphere, at the bottom of the Mariana's Trench, in volcanoes, in oil wells ten miles deep, and even in nuclear reactors.

So voids don't exist for very long in nature wherever life is capable of growing. That means whether you're talking about soil or the microbiome of the human gut, something will quickly come in to fill any space that's fertile enough to host life. And it will continue to grow until the space is completely full.

It's for these reasons pathogens can only take root when probiotic organisms are absent. You have a certain amount of available space in various places throughout the body. And that space is going to be colonized by one life form or another – for better or for worse.

When beneficial critters are present in full force, and at their full strength, pathogens set up shop, but their numbers are kept in-check. On the other hand, when the good guys are depleted or weak for some reason, then, and only then, can pathogens take over and cause you problems.

Sunlight supports life. Moonlight supports death and decay

Sunlight and moonlight have distinctly different qualities. Sunlight gives us (used to give us) golden light that's warm, drying, life-giving, and topically antiseptic (cleansing). Our bodies need it to make vitamin D, which is essential for dozens of bodily processes.

Sunlight helps regulate our pineal gland and the production of a cascade of all-important biochemicals, including tryptophan, 5-HTP, serotonin, and melatonin. Our sleep cycles, our hormone levels, and our mood depend on sunlight.

Plants also get their food/energy from sun. Plants use sunlight to make chlorophyll. They need its warmth to keep down microbial populations on their surfaces. And these foundational processes support all life on earth by feeding the animals above them on the food chain (and so on). So sunlight is kind of important.

On the flip side of this yin-yang duality, the moon gives off a silvery light that's cold and dampness-forming. Believe it or not, the moon's rays have actually been shown to have the opposite effect of sunlight:

They actually reduce the surface temperature of objects nearly 2 degrees Fahrenheit when exposed to direct moonlight vs. "moonshade", and considerably more when concentrated with a magnifying glass. These effects wouldn't happen if the moon merely reflected sunlight.

So the moon promotes necrosis (cell death), sepsis (self-injury through the body's immune response), and putrefaction (rotting)

Night is when the molds and fungi (and some bacteria) come out to play. The nighttime is necessary for the microbes in charge of decomposition to break down leaves, fallen branches, diseased plants, and dead animals. This decaying matter decomposes and feeds the entire food chain after it's recycled back to the earth.

And it's for these reasons the sun and the moon have been used as a metaphor for the battle between good and evil through the ages. It's the reason ancient myths and legends told tales of astrological gods that ruled the sun and moon for millennia.

Far from mere entertainment, the practical observations and applications of sunlight and moonlight have fascinated man from ancient folklore to the present day for very good reason.

Yet we tend to look back at what ancient man understood, and who he honored/worshipped, and we think of them as amusing customs and primitive beliefs peculiar to their tribe. Little did we know they had more basis in truth than we ever gave them credit for. Hmmm… seems to be a recurring theme throughout our belief structures and approaches today.

To sum it up, there's far more value in traditional practices than we thought, while our modern methods are rife with flaws that are coming back to haunt us in ways we never imagined.

Our very way of life (and those who design it) now promotes death and decay

Who can deny, we're being pushed more and more toward sepsis, necrosis, and degradation. Our culture is being driven into breakdown mode by almost everything in our environments – low quality food, contaminated water, polluted air, dirty sources of energy, lifestyle choices (like cigarettes and alcohol), medicines, electronic gadgetry, and the EMF waves they produce.

It's plain theft of our life-giving energy and our resources. This is causing us to live in a perpetual state of survival mode or decline, rather than our natural state of being, which is to expand and thrive. Just look at plummeting fertility rates for proof.

And it's not just people. Look closely at the plants in your neighborhood – trees, shrubs, and grasses. Do they look like they're thriving? For those of you old enough to remember the days before cell phones and Wi-Fi, do you see as many insects today as you used to (that feed the food chain)? Do you still see early morning dew on plants anymore?

These are just a few effects you might see around you. Do a quick image search on the Internet for "massive fish die off" and "massive dead birds." Read up on how fast the Great Barrier Reef off Australia is perishing (because almost all of its 133,000 square miles is dead).

After you do those things, ponder this: The things that are stressing plants, animals, and ecosystems are affecting you and your biochemistry the same way it's affecting them right now. The human body just has more sophisticated coping mechanisms than lower life forms enabling it to be more resistant to overt states of disease.

In other words, The Powers That Be are pacifying you and me with all sorts of chemical, pharmaceutical, dietary, lifestyle, and psychological entertainment, while their products, programs, and approaches promote disease, degradation, and death. They're killing us slowing but surely.

The circle of life on earth is supposed to be balanced between growth and decay. In fact, it would be fair to say birth, growth and life are supposed to be the favored, more prevalent, state of nature. But there's way too much degradation woven into our society now, with more being added at an alarming rate. Disease rates testify to this.

And, on a planetary scale, it's quite obvious the entire earth is finally beginning to show irrefutable signs that it's suffering from an advanced state of disease. A perpetual state of struggle and survival is the new normal. That seems to be the agenda going on right now. And you know what? It's working.

Myths: The human body is stupid and fragile. It doesn't know what it's doing. It malfunctions easily. And it needs man's help to run well

The reality is, quite the opposite is true: The human body is exquisitely engineered to be unbelievably intelligent, and shockingly robust in adapting to environmental challenges.

Your body is so staggeringly complex that science and medicine, for hundreds of years, have failed to recognize or appreciate what it's doing when it's either functioning properly, or it's trying to fix a problem situation with the tools it has at hand.

It just takes a beating and keeps on ticking. And it can take decades of reckless, completely irresponsible, abuse before it gives up and sends you a memo in the form of symptoms that say, "No more... Unless you make some changes, things are going to get ugly."

Unfortunately, we as a culture never learned how the body works, why it malfunctions, what the symptoms mean, or what to do about it (at least not the truth). Our institutions have dropped the ball in determining the true nature of how the body works, and educating us about these things.

And so they've left it up to us to seek out this scattered knowledge, to separate fact from fiction, and to fight The Establishment to get what we need. For example, did you know **most health problems are present in the body for years or decades before you notice significant symptoms?**

Your body can compensate and fight its way through an imbalance or health threat for quite a long time before it's finally had enough and can't take it anymore. It's only after the body's compensation mechanisms are exhausted and broken down that disease starts to win the battle and symptoms start to appear.

In fact, most organs and processes in the body have lost over 75% of their functionality before they start complaining and giving you noticeable symptoms

Before that, you wouldn't even notice your body is fighting all sorts of environmental threats all day, every day. This is the way most imbalances progress into diseases, because the body, in its rugged engineering, doesn't want to trouble you with every deficiency it's trying to manage, and every health challenge it's trying to overcome.

It just goes about its business and doesn't complain very much. It's only when you – the manager of your own health – are so far in arrears with your nutrition, detox, and lifestyle duties that the body calls "foul" and initiates emergency proceedings.

It's only after the over-taxation of organs turn into blockages and breakdowns that the body calls '911' and asks for help by giving you observable, noticeable symptoms that are so bad you have to stop, take notice, and hopefully change your ways. It's only when your healing mechanisms are in pretty bad shape that you get symptoms that compromise your lifestyle, or threaten your life.

In the meantime, the most your body will ever trouble you with is a rash, a nagging cough, a little numbness in the extremities, a tummy ache, a mild pain that comes and goes, or feeling/behaving unusual for no apparent reason. These are all signs the body is trying to deal with some sort of deficiency, or it's actively engaged in some sort of healing process.

For instance, you can have lots of cancer cells growing in your body for years and be completely oblivious to it because your body is constantly killing worn out/damaged cells and replacing them with new ones. With 50 and 70 billion cells being replaced every day, the removal of cancer cells would go unnoticed.

At the same time, you can consume hundreds of times more sugar than your body was designed to handle. Yet the majority of people will never know the stress they're putting their pancreas and cells through until they start to show symptoms of diabetes and metabolic syndrome.

And, germane to this discussion, you can assault your gut flora with a long list of chemical and biological assailants, for generations as it turns out, and the microbiome still manages to bounce back and do the best it can. Unfortunately in the case of gut flora, by the time you notice any symptoms, your microbiome is likely to be quite corrupted.

So, as a general rule, when your body is healing something, you want to err on the side of leaving it alone and letting it do its thing. Don't intervene and try to fix the symptoms, as conventional medicine is so fond of doing.

That frequently may give you some relief in the short run, but makes things worse in the long run, because those symptoms are the body's call for help. In short, you want to figure out what caused the problem and solve that, which is now possible in almost every case, when you know where to look, and who to listen to.

Symptoms are not the source of health problems. They're a red flag... a key to decoding the problem and finding a solution... if you listen carefully to what they're saying

Symptoms are not the source of health problems, as organized medicine would have us believe. Rather, they're the "tangible" expression of imbalances within the body that lead to states of dis-ease. They're a feedback mechanism designed to make us aware of how stressed or damaged the body is, now showing up to warn us about our iffy ways.

In the early stages of chronic disease, symptoms warn you about deficiencies, imbalances, and malfunctions it's facing and needs your help to deal with. You can think of them as warning signs to change your ways while you're still able to do something about it. Then, when they get to an advanced stage, symptoms are a virtual cry for help when something upsets the body beyond its ability to compensate.

An example we're all familiar with: Pain. Pain isn't the source of our problems. It's simply the thing that bothers us… and the manifestation we desperately want to get rid of. But it's the splinter, the broken bone, the arthritis, or the blocked artery in the heart that's really causing the pain. And that's what we should be focusing our efforts on fixing – the source of imbalance and disease – once we know the truth about where chronic disease comes from.

Which brings us to one of the main points I hope you to take away from *Gut-Brain Secrets*. **The cause of most chronic diseases is not a mystery that researchers will one day discover many decades, thousands of man-hours, and billions of dollars into the future.**

Rather, it's already known to those who are willing and able to know the truth when they hear it. Moreover, solutions to reverse most diseases are also understood, partly or completely, by those same people.

Let's put this in no uncertain terms: For over a hundred years, Western medicine has made its living off of keeping people in the dark about what causes disease. And to this day, it continually has people "running around in circles" trying to fix one symptom after another with drugs, palliative treatments, surgeries, and partial solutions, while failing to address, or even acknowledge, the true root of the problem.

It has people of the trusting public fixated on covering up symptoms – often with drugs, often perpetuating the root cause, and all-too-often creating side effects that need to be dealt with. Object and deny as they might, that's the way Western medicine operates at all levels.

The unconscionable part is, the Medical Establishment spends billions pretending to work day and night for decades on end "racing for the cure" (you name the disease). But it's a façade.

They don't want treatments that are too cheap or too effective. That's bad for business.

Rather, they want slow, weak, expensive, partial solutions that work decently well for a good percentage of people, most of the time. In other words, they want us all to be in a state of partial health/partial sickness. And, most important, they want everyone dependent on them.

They want to control you. So every time a highly effective treatment, or, heaven forbid, a cure emerges, they stamp it out one way or another as if their income stream and power over the people depended on it.

Your call to action: If you want be in excellent health for your maximum intended lifespan, get educated about what causes imbalance and disease to happen in the body, and what causes optimal health to take place. Understand the basic concepts about how the body works and the hazards threatening your health all around you.

Learn what your symptoms are trying to tell you about what's happening in your body, and you'll be halfway home in reversing them before they cause a regrettable situation. And, most important, don't blindly believe any source, no matter how credible they're held out to be. Instead, judge for yourself the mechanisms of action, the rationale behind the claims, and the 'why' in every circumstance.

We know more about the body than we've ever known. We have more options than we've ever had. And now your family's health depends on you knowing the right information.

Bottom line: Your body is infinitely smarter at healing itself than any doctor, healing modality, program/system, or organization

If you understand what your symptoms are telling you, they can actually inform you and empower you to make better choices. Symptoms can be your friend, or your enemy, depending on how you interpret their meaning.

If you use drugs and mainstream medical interventions to continually suppress symptoms, the symptoms may persist and get worse. On the other hand, if you find out what's causing the symptoms and fix that instead, then symptoms can be just an early warning sign and a blessing in disguise. It all depends on your health & nutrition literacy, and your level of responsibility in seeking truth and taking action.

The immune system employs pathogens to chelate heavy metals

Believe it or not, the human body uses viruses and fungi to do its bidding in the war against toxins. "The enemy of my enemy is my friend." Buried deep within the infinite intelligence of the human body, the immune system can actually increase or perpetuate infection activity of pathogens as a way to chelate otherwise hard-to-handle stores of heavy metal and chemicals.

In some cases, the body will allow an active viral, bacterial, or yeast infection to persist longer than usual because that germ is detoxing heavy metals that would otherwise be challenging for the innate detox system to remove. While in other cases, it may let an old virus out of storage somewhere in the body and revive it in order to combat a particularly dangerous threat.

The theory behind this is, different types of microorganisms replicate and operate in different ways. Some of them have special access or ability to chelate heavy metals that normal immune cells and processes don't have. Some of them just like the company of heavy metals.

So when you interfere with this counter-intuitive healing process by taking antimicrobial/antifungal drugs, reducing a fever, or otherwise cutting a sickness short, you may be undermining the body's ability to eliminate toxins.

In other words, sometimes a never-ending illness or overgrowth is the body's backup strategy to manage an on-going, more destructive, source of heavy metal poisoning that it can't handle through conventional means. Likewise, many people believe the immune system can employ less destructive pathogens to fight more destructive ones.

One example specific to mercury chelation: Many practitioners suspect the body will keep candida infections going for prolonged periods in an effort to help remove large, on-going influxes of mercury from big fish or dental amalgams. And for that reason a candida infection will be extra hard to extinguish until you fix the mercury problem first.

How hunger cravings work

Your body has complex nutritional needs that change from moment to moment. One minute you may be stressed at work and need extra anti-oxidants. Ten minutes later you may be exercising so you need extra potassium. After working out you may need extra protein to rebuild muscle tissue.

At any given moment there are dozens of processes happening in the body, and many dozens of nutrients the body needs to run those processes. The amazing unseen mystery within is that the body has to communicate those needs to your brain and taste buds on an unconscious level to compel you to eat what it needs.

Think about it: If your body needs X amount of vitamin D, Y amount of protein, and Z amount of magnesium, how is it possibly going to tell you the specific nutrients and exact quantities it needs? And then how would your rational mind figure out which foods you need to eat, and how much of each you need to eat, in order to get exactly the right proportions of nutrients to satisfy that laundry list of demands?

This is another instance where the body's innate intelligence is nothing short of genius. All the systems in your body – your gut flora, the cells of your gut wall, systemic cells, hormones, neurotransmitters, and nervous system – they all work in concert to tell your brain and taste buds what to eat. The end result being hunger cravings. Here's the idea:

First, your body senses it's low on certain vitamins and minerals and needs replenishing. Through miracle of Nature and experience, your body knows what nutrients are found in naturally-occurring animal products, plants, simple ingredients, and seasonings that you've eaten before.

It knows which foods have which vitamins, minerals, enzymes, and other nutrients in them. In fact, your gut knows better than your conscious mind what's in the food you eat. It knows that oranges give you Vitamin C, carrots have Vitamin A, and steak is an abundant source of protein.

So it's keenly aware on a subconscious level which foods to ask for to get the nutrients it needs. And that's one reason you may crave a grilled cheese sandwich when you need calcium, fresh berries for your fruit salad when you need antioxidants, or fish when you need omega oils.

That's how hunger cravings work. Your body is constantly monitoring nutrient levels and adjusting its hunger cravings according to what your body needs at that moment. It even requests any number of over 72 micro-nutrient trace minerals by craving sauces, ingredients, seasonings or spices like salt.

Of course with the way things are, the body doesn't always get the nutrients it asked for, and it needs to trigger hunger sooner than it would, had those nutrients been received.

But, just to be clear, this phenomenon applies mostly to foods that haven't been modified all that much from their original source material. You can disregard everything just mentioned when it comes to commercial "fake foods" (i.e. glutamate-based flavor enhancers, artificially flavored, highly processed, or otherwise re-engineered by man).

They don't count because industry is great at tricking the tongue-brain system with a wide variety of nose and tongue manipulation, thus disturbing normal hunger craving and nutrition patterns.

Think the body has no use for the appendix? Think again

Medical experts concluded decades ago the appendix may have once served a useful purpose in man's ancient past, but we've evolved out of it since then. They said it's so utterly useless that when a person goes in for lower abdominal surgery, surgeons will often remove the appendix just to keep it from causing problems down the road.

Imagine that: Doctors haven't figure out what the appendix does. So they conclude it does nothing and remove it just for the hell of it. Unfortunately, this is another case where the experts got it wrong.

The appendix actually does serve a purpose – more than one purpose actually. It's not some evolutionary accident or obsolescence. The appendix is actually a reservoir for storing "replacement" bacteria. And the body uses it to repopulate the gut after a major assault to the microbiome, like dysentery.

However, it can only distribute what it's got inside. So if you've got a Corrupted Gut, your appendix probably suffers from the same unhealthful bacterial profile as your intestinal tract. Maybe that's why appendixes get infected and rupture? Makes too much sense, doesn't it?

If all this is true it's another reason why reversing a Corrupted Gut can be so challenging – the appendix keeps the status quo going, for better, or for worse. And probiotic supplements can't repopulate the appendix directly. They only have access via the gut.

"Man's Way" vs. "Nature's Way" ... That is the question

To be sure, each approach – the medical/interventional approach vs. self-healing – has its advantages and disadvantages. Each has its place in healing.

To name a few, Western medicine is great at keeping you alive in life-threatening emergencies. Mainstream medical care is great at trauma care. Nature's Way, not nearly as much. On the other hand, Western medicine sucks at explaining, preventing, and, some would say treating, chronic, degenerative diseases. While Nature's Way has far better answers for healing thyself.

The big takeaway from this health information saga is that we should start forming our lifestyle choices and treatment decisions around accurate thinking. We need to seek to understand where our problems are coming from first, so we're not continually butting our heads up against a wall trying to solve the wrong problems, as modern medicine has a habit of doing.

The lesson to learn here is, our approach to nutrition, disease, and wellness desperately lacks knowledge and responsibility today. And that's what we need to change if we want to improve our health outcomes from 49th in the world as of 2017, to where we deserve to be, considering the U.S. spends far more than any other country in the world on its healthcare.

That's where we need to start from. Knowledge and personal responsibility are what will save us from the society-wide collapse in mental and physical health we're experiencing right now, not to mention save our healthcare system from financial ruin.

CHAPTER 4:

Redox Signaling Molecules and Inflammation

(How Oxidation, Reduction, and Inflammation Heal or Harm the Body)

(Based on the work of Gary Samuelson, PhD, and Dr. Zach Bush)

4

Ever wonder how 30 trillion *human cells* and 100 trillion *microbiota* live together harmoniously as one human being?

Answer is, *communication* lets trillions of microscopic cells (e.g. your cells, friendly bacteria, and pathogens) live together as one macroscopic organism (e.g. you, animals, plants). It's re ally quite remarkable when you think about it, because without incredibly well-designed communication and control systems keeping all the members of the choir on the same page, singing the same tune, it would be utter chaos on so many levels.

Just imagine what would happen if each cell selfishly did whatever it wanted, whenever it wanted. Imagine what would happen if cells called for help and their calls went unanswered when under attack, when needing repair, when malnourished, or when their DNA's been damaged.

That's the difference communication makes in ensuring the body's needs are being met at all levels – whether that be at the cell level, organ level, organ system, or complete organism.

In the same way that many channels of communication are needed to run households, cities, nations, and organizations effectively, many channels of communication are also needed to orchestrate the body's essential functions.

Clear lines of communication are needed to relay messages between parties, protect and care for its citizens, replace damaged facilities, allocate resources appropriately, and control activities that benefit the group. Because without cooperation and coordination at many levels, complex organizational structures – like those found in animals, plants and planets – would not be possible.

More than anything, that basically means having a variety of ways one member of a community can get a message to other interested parties, be they fellow members at the same level, or "governmental" response teams higher up the chain of command. In a word, it's *communication* – or *signaling*, to be precise – that turns many individuals into one smooth running (healthy) organism.

To name three channels of communication in the human body, you know about DNA, hormones, and neurotransmitters:

1. **DNA** transmits the organism's genetic information (source) into the construction (coding sequence) of proteins and cells.
2. **Hormones** regulate the activity of organs at a distance via messenger molecules sent through the bloodstream. And
3. **Neurotransmitters** conduct signals along nerve pathways using electro-chemical signaling molecules such as dopamine and serotonin.

Redox signaling

Just as important to our physiology, there's one communication system most people know nothing about. It's so new to science most people in the general population have never even heard of it, despite the fact it's essential to all life on earth. And that system is the network by which communication happens at a cellular level, called redox signaling.

Conveyed through compounds called redox molecules, mitochondria (the power plants of cells) make one type of redox molecule – called *oxygen* redox molecules. And bacteria make another type – called *carbon* redox molecules. Each makes its own type of redox molecules for its own purposes, thus forming two distinct communication networks.

Mitochondria make and use oxygen redox molecules for signaling among human cells, while bacteria make and use carbon redox molecules for signaling between each other and cells of the gut. We'll start the discussion that follows referring mostly to mitochondrial redox molecules, and talk specifically about bacterial redox molecules later.

Mitochondria in cells use their redox signaling to

- **Protect** against damage from foreign invaders and toxins,
- **Detect** damage after it happens, and tell the immune system the extent of the damage,
- **Repair** cells when the damage is reversible, and
- **Replace** cells when the damage is too severe to repair.
- **And a whole lot more.**

Their microscopic counterparts, bacteria, use their carbon redox molecules to sense the state of the microbiome and the gut lining. It's through their redox signaling network bacterial species talk to each other, as well as organ systems, so the various parties can put their healing mechanisms to work fixing imbalances and dysfunctions.

Redox signaling is a huge deal because it's through this communication system that the various parties can interpret the condition of the situation correctly inside and outside cells, and can then do something about it.

Conversely, when communication breaks down, no one can read the situation correctly, and health challenges happen or are allowed to continue, such as stress, microbial attack, physical injury, toxicity, malnutrition, oxygen deprivation, or DNA damage.

This new science of redox signaling is a massive breakthrough in our understanding of how dysfunction and disease materializes. And the advances of just the last 5-10 years represent a giant leap forward in reversing them.

As one of the leading researchers in the field, Gary Samuelson, says, "If even 1/10 of the promise of redox signaling plays out, it will revolutionize functional medicine and nutrition as we know them." His colleague, and another leader in the field, Dr. Zach Bush, believes successfully stabilizing redox molecules outside the body could be one of the most important discoveries in human history.

Oxidation and reduction overview

Over the last two or three decades, you may have heard marketers and wellness coaches talk about the influence of free radicals and antioxidants on our health and longevity. They're elements of the "oxidation-reduction" cycle in the rapidly emerging science of Oxidative Medicine, or Oxidative Sciences – now called "redox" in many instances.

Redox reaction with electron transfer

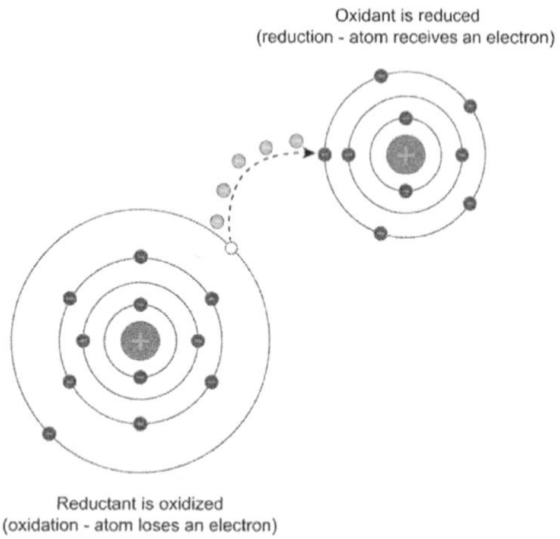

Oxidation-reduction: Simply put, oxidation and reduction describe the exchange of electrons.

Electrical charge: Electrons are negatively charged. So *removing* electrons through oxidation increases positive charge and acidity, while *adding* electrons through reduction/antioxidant activity does the opposite – it increases negative charge and alkalinity. This is important because most pathogens, toxins, and free radicals are in their comfort zone when positively charged and acidic, while antioxidant activity fights those threats by donating electrons, thus increasing a molecule's negative charge and alkalinity.

Oxidation is the "stealing" of electrons from a molecule – or an *increase* in the state of oxidation. So molecules with the propensity to oxidize substances are called "oxidants." To illustrate, when oxygen takes electrons slowly from iron, that's a form of oxidation we call rust. When a flammable material burns or explodes – again, with oxygen – that's oxidation happening rapidly right before your eyes.

Oxidants, aka reactive oxygen species: In biological systems, oxidation destabilizes the matter inside cells by stealing their electrons, which makes them blow apart in a hurry unless they find an electron to pair up with and zero out their charge. For that reason, oxidants are used by our immune systems as the ultimate antimicrobial, detoxifying agent.

Most oxidants, as the name would suggest, are predominantly oxygen-based molecules, so they're called "reactive oxygen species" (ROS). Nitrogen and sulfur also form their own less-common reactive species. Oxygen, hydrogen peroxide, ozone, and chlorine dioxide are some of the best-known oxidants in the functional medicine field.

Free radicals: When a molecule with a balanced pair of electrons loses one of those electrons due to oxidation, the resulting molecule can become a "free radical" because of its extra, unpaired electron, which is highly reactive and potentially damaging to cells. Its destructive effect on cells is why the public was taught to fear free radicals in the 1980's-2000's, and why we were instructed to get plenty of antioxidants to combat free radical damage.

Reduction is the opposite of oxidation. It's the giving of electrons, or a *decrease* in the state of oxidation (hence the term "reduction").

Reductants: Molecules that give up their extra electron in chemical reactions are called "reductants" (even though that sounds backwards), or Reduced Species (RS).

Antioxidants are tiny molecular catalysts that make oxidants give their extra electron(s) to reductants, thereby neutralizing (balancing) them both of electrical charge and biological reactivity. The body's naturally-produced antioxidants such glutathione can perform tens of millions of these reactions per minute. Reactive oxygen species and reductants then turn back into salt water (from which they came).

Redox: Not too long ago, redox scientists realized that saying the words *oxidation, reduction,* and *reactive oxygen species* out loud made them sound like nerdy chemists who use big words to confuse or impress people. So they came up with a cooler sounding umbrella term to describe the players and processes (e.g. redox molecules, redox reactions).

The field swapped and shortened oxidation-reduction to "redox," which is short for REDuction-OXidation. And it seems to have a ring that's right for the mainstream.

Redox molecules: Reactive oxygen species (ROS), and reductants/reduced species (RS), are collectively called "redox molecules" or "redox signaling molecules." Redox molecules are by-products of metabolism that (1) mitochondria use to support cells in many ways, or that (2) bacteria use to support the microbiome.

Mitochondrial redox molecules: Without the concentrated heat of a conventional fire, mitochondria burn fat or sugar in the presence of oxygen to make ATP to fuel cells. This produces oxygen redox molecules as a by-product.

Otherwise known as "metabolism," aerobic exercise dramatically increases the need for, and rate of, this process. Made principally of oxygen, mitochondria's *oxygen redox molecules* are the communication network between mitochondria and human cells.

Bacterial redox molecules: When bacteria metabolize food, they spit out their own variety of redox molecules as a by-product. Made principally of carbon with about 17 potential binding sites (representing its signaling capacity), each of the tens of thousands of bacterial species makes about 10-15 different varieties of these carbon-based redox molecules.

Oxidative stress: Oxidative stress is the amount and duration in which oxidants outnumber reductants. Oxidative stress can be damaging when insufficient antioxidants and reductants are available to neutralize the oxidants – particularly, in chronic, uncontrolled circumstances. Oxidative stress can be beneficial when used therapeutically.

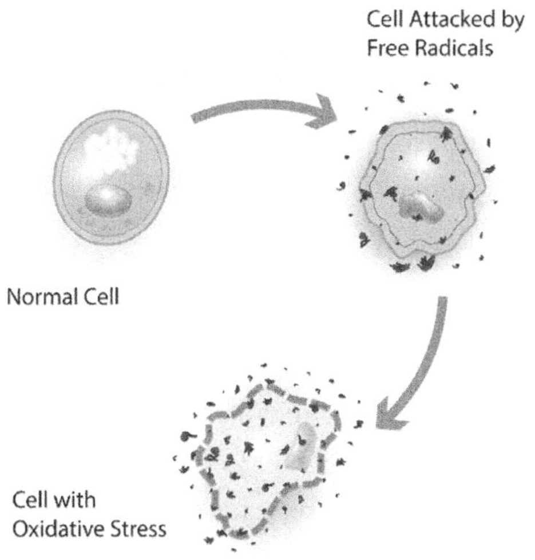

Tight junctions are the filaments that normally hold the cells of all our membranes together and keep unwanted things out. They also open and close on-demand to let authorized things through. However, when tight junctions become damaged, they stay open and let unauthorized substances through, causing myriad health problems.

Why oxidation and reduction matter

Oxidation and reduction are important to human health because oxidation-reduction cycling – the giving and taking of electrons – happens in all life on earth and is a fundamental part of every biochemical reaction.

Specifically, oxidation and reduction reactions control youth, healing, energy production, immunity, detoxification, hormonal response, and support of the microbiome – when it's working properly.

And it causes a lack of cell repair, a long list of chronic, degenerative problems, and accelerated aging when it's not. So it's more than important to all life on earth. It's essential. Life would cease to exist very quickly without oxidation and reduction.

Oxidation and reduction, or redox signaling (they're the same system), is the biggest, most important story for medical science to research today, and you to learn about, in our on-going effort to get healthy and stay healthy. It's right up there with the gut-brain connection.

In fact, the two phenomena, the gut-brain connection and cell-cell communication, are partners in corrupting our health or making us well. They're inseparable links in the same chain of events that manifests as ADD, GAPS, and a truly staggering array of diseases and dysfunctions skyrocketing in every statistical category and population group on earth.

Redox reactions are the basis of all communication, healing, and repair when it's working (i.e. balanced)… many break downs and failures when it's not

- Metabolism/energy production.
- Healing at cell level.
- Communication to and from cells (redox signaling).

- Defense, antimicrobial.
- Proliferation of mitochondria (which make energy and redox molecules).
- Increase diversity of microbiome – particularly rare species of beneficial bacteria.
- Suppress pathogens in microbiome.

Negative charge and alkalinity are synonymous with health and healing, while positive charge and acidity creates inflammation, disorder, and disease

If you had to pick only one measurement to assess your relative state of health or sickness, a person that knows their stuff would pick electrical charge and pH. You can think of the two as pretty much the same thing in the body, because they're like twins or dance partners.

When oxidation steals electrons (which are negatively charged), cells and atoms become positively charged. That means greater instability, increased reactivity, acidity, corrosion, and inflammation that doesn't know when to quit. It means your cells and atoms are literally falling apart and not repairing themselves efficiently, which is imbalance, sickness, and aging, in a nutshell.

On the other hand, when your immune system is good at making antioxidants, your cells and atoms have all the electrons they need to replace those that are lost through oxidation. We call this neutralizing oxidative stress. And that equates to greater negative charge, higher alkalinity, better cell-cell communication, inflammation that starts and stops appropriately (aka acute inflammation), and healing capacity to spare.

Intertwined with all that, acids have more protons than electrons, which gives them a positive charge. The more extra protons a substance has, the more acidic it is. And the opposite is true – the more extra electrons something has in relation to positively charged protons, the greater its alkalinity. Hence, many alkaline foods and substances are famous for their healing/antioxidant effects.

In short, extra electrons are *negative charge* and *healing capacity* through oxidation-quenching and helping the immune system hear when inflammation needs to be shut down (alkalinity supports cell-cell communication). On the flip side, electron deficiency is *positive charge* and *poor healing ability* through corrosion and failure to realize inflammation is not working the way it should. Acidity aids and abets that.

We used to think free radicals were harmful by-products of exercise

Until just a few years ago, pretty much everyone in the wellness field thought free radical oxidation harmed cells and accelerated the aging process … and that was all there was to it.

It was part of this whole "exercise-induced free radical damage and antioxidant protection" paradigm pretty much everyone was preaching at the time. Researchers and educators basically thought aerobic exercise and cellular energy production make reactive oxygen species by accident.

Because oxidants are very good at destroying things, they believed the resulting free radicals were responsible for randomly injuring cells and making us age before our time. Therefore, we need to protect ourselves against this sort of cell damage by eating foods high in antioxidants.

But science has since come to its senses as research has revealed reactive oxygen species (ROS) are not harmful by-products of metabolism that you need to eliminate, else they harm cells and cause premature aging. Instead, they're made for a very good reason – many beneficial reasons, actually.

Researchers have come to realize the body uses oxidants in a controlled manner to selectively repair or replace injured cells, depending on the extent of the damage. Even more pertinent to this discussion, mitochondria make redox molecules to send signals about what's happening inside the cell.

So redox molecules are both the communication network of the mitochondria, and they're the disinfectant that wipes out unwanted material to make way for healthy, new cells.

But up until just 5-10 years ago, scientists didn't realize that a healthy system purposely makes an equal amount of destructive oxidants and neutralizing reductants to maintain balance. In other words, good health is a balanced redox system.

In fact, large amounts of both ROS and RS are made when you exercise. So, oddly enough, taking a redox molecule supplement does many of the same things that exercise does; whereas, ingesting antioxidants right before you exercise negates much of its benefit, because the cleansing/restorative benefit of exercise gets negated.

We used to think antioxidants just came from food

We now know our own cells make large quantities of antioxidants that are far superior in reach and capacity to those supplied by foods. In fact, most antioxidants that come from foods can't get inside cells, so they don't do what scientists thought they did (Vitamin C is an exception).

That said, they do offer some benefit quenching free radicals in the blood and extra-cellular matrix. For instance, alkalizing foods (and substances) such as raw vegetables contain an abundance of negatively-charged hydrogen ions. These act as electron donors (i.e. antioxidant effect). And that's one of the primary reasons raw, naturally-grown vegetables are so good for you.

Therefore, antioxidants from food are hugely beneficial. They're just not as all-powerful as those the body makes internally. To give you some idea, dietary sources of antioxidants contribute about 4% of the body's total supply, while internally produced glutathione by itself represents about 85% (and is far stronger).

Our new understanding of oxidation and reduction

We now know oxidation and reduction is a carefully orchestrated, essential process the body uses to protect itself from threats and clear up damage, so that repair and renewal processes have a pristine environment in which to rebuild.

It's fundamental to human life because the body uses oxidation to destroy everything it deems undesirable – microbial pathogens, toxins, heavy metals, dead cells, and unrepairable cells. Without oxidation and reduction, you'd have mere hours to live … a day or two max.

However, oxidants are like wild beasts in that they're useful (crucial, really) when kept under control, damaging and problematic when they get out of control. Therefore, when inflammation is done blowing its target to bits with oxidants, antioxidants bring unused oxidants together with reductants and make them exchange electrons.

This neutralizes the oxidant of its destructive capacity, and turns both of them back into harmless salt water. The body's own antioxidants like glutathione can do this tens of millions of times per minute.

James Watson, PhD. (Nobel Prize Laureate who discovered the structure of DNA with Francis Crick, PhD.):

> *"In the chemistry of oxidation and reduction reactions, the body cannot survive without making both oxidants and antioxidants. There is a delicate balance between the two. Physical exercise prompts the body to make large numbers of oxidants – molecules called reactive oxygen species, or ROS."*

Exercise benefits the body by enhancing oxidation, reduction, and redox signaling

Exercise is oxidative stress. Oxidative stress is exercise. At the same time, the ability to sustain exercise is determined by mitochondrial energy production capacity, antioxidant capacity, and balance between oxidation and reduction.

First and foremost, exercise is an oxidative activity. It's controlled oxidation. Meaning, exercise produces a cleansing and renewal effect at a cell level. That's *why* it's good for you, and *how* it benefits the body. It leads to very important mechanisms of cell regeneration, exercise capacity, and speed of recovery.

You see, exercise makes mitochondria burn fatty acids or sugar along with oxygen to power cells. That mitochondrial metabolism produces large amounts of oxygen redox molecules that act as both universal cleansing agents (that need to be neutralized) and message carriers.

Exercise, oxygenation, and resulting oxidation capacity also help regulate electrical charge (i.e. lots of negative charge and antioxidant activity), pH balance (deep breathing is alkalizing because it releases carbon dioxide, which acidifies the blood), and cell wall permeability through greater charge (i.e. better hydration, nutrient exchange, and detoxification).

As we discussed in the previous chapter "The Wisdom of Nature vs. the Fallibility of Man," those are several key factors that promote either a high state of health and healing when they're present, or disease conditions that favor pathogenic activity and the breakdown of the body when they're absent.

Point is, exercise, oxygenation, and oxidative activity are precious to your wellness. But the other half of the story is, it takes time to build up a capacity for exercise, which is another way of saying *antioxidant capacity,* which completes the oxidation-reduction cycle. Otherwise, you experience extended recovery time, unrepaired damage, and overuse injuries when you exercise too hard when you're out of shape.

Through redox molecules (reactive oxygen species and reduced species), exercise dramatically increases protection of cells, healing of cells, balance, and a laundry list of good things. When well-controlled, oxidative reactions represent most of the body's mechanisms of youth and healing.

Conversely, when those systems are overwhelmed or fail altogether (i.e. ROS not being neutralized properly), oxidative damage is the mechanism of imbalances, exposure to genetic weaknesses, breakdowns, and rapid aging.

To sum it up, exercise is oxidation. Exercise is cleansing and rejuvenation on a cell level. But exercise needs to be matched up with an equal amount of aerobic fitness, which is determined by your mitochondrial energy production capacity, antioxidant/reduction capability, and balance between oxidation and reduction.

You need all that to sustain aerobic exercise. Otherwise, you get too much breakdown, and not enough repair and replacement. So basically, aerobic capacity is brought about by a redox signaling system that performs well under load.

Cells communicate and heal with redox molecules

Like a vast military operation with a lot of moving parts, both mitochondria and bacteria in the microbiome need excellent communication to perform efficiently and effectively under all conditions. Redox molecules are that communication network enabling cells to talk to each other and "command central," thereby allowing immune cell armies to do their jobs.

Through redox signaling, the body is able to ...

- detect when cells are under stress,
- tell DNA to activate coping mechanisms,
- call in the immune system to fight off threats,
- repair mildly damaged cells,
- hit the self-destruct button on badly damaged cells,
- regulate hormonal response,
- regulate mitochondrial metabolism,
- turn off coping mechanisms after a threat subsides,
- plus a whole lot more.

To illustrate how redox molecules work, when cells become damaged for any reason – toxicity, infection, physical injury, lack of oxygen, or even malnutrition – they enter a state of oxidative stress, which means more reactive oxygen species are supplied to the area than there are antioxidants and reductants to neutralize them.

High amounts of reactive oxygen species are the "S.O.S. distress signal" that tell cells and systems that there's a problem that needs to be addressed, where it's coming from, and how dire the situation is.

Like smoke coming out of a burning building, the more oxidant molecules the immune system finds leaking out of cells (because they were never neutralized by antioxidants/reductants), the more emphatic the distress signal is deemed to be.

Later, when the situation has been resolved, the lack of free-floating redox molecules inside and outside cells prompts nuclei and the immune system to turn off those coping mechanisms and go back to business as usual (i.e. homeostatic balance).

So it's through mitochondrial redox signaling that the nucleus "reads" the condition of oxidative stress occurring in the cell and can activate a variety of genetically-controlled coping mechanisms to deal with the threat and restore balance to the system.

Of course, the majority of the time a cell reads its status report and realizes it's fine and doesn't need repair. Still other times, cells take a look around them and realize they're different from that of the host organism. They realize they're a cancer cell and need to be sacrificed to protect the health of the whole.

But the thing is, without a large vocabulary with which to communicate, cells don't even realize what a healthy cell is supposed to look like, versus what a cancer cell looks like. They don't talk to each other fluently about their status and their needs. And they don't relay that information to the organ systems that need to know such things, such as the immune system.

Summary: In the same way that analyzing a car's exhaust tells a car mechanic how well an engine is running, the ratio and volume of ROS to RS indicates whether the cell is happy or distressed. Mitchondria, cell nuclei, and organ systems can then use this information to activate healing, or turn it off.

Here are some of the "buttons" the nucleus can push to call for help

1. **The DNA Repair** "button" mobilizes the DNA damage detection and repair crew.
2. **The Antioxidant Boost** button makes more antioxidants to neutralize the potentially harmful surplus of oxidants.
3. **The Intercellular Communication** button strengthens lines of communication between cells.

4. **The Increase Blood Supply** button dilates blood vessels to cells that could use more resources.
5. **The Stronger Cell Adhesion** button makes cells hold more tightly to each other.
6. **The Inflame Tissue** button stops damage from spreading.
7. **The Secrete Antibiotics** button deploys antibacterial substances to fight foreign invaders.
8. **The Stop Cell Division** button prevents distressed/damaged cells from replicating.
9. **The Send Distress Call** button sends a distress signal to the immune system.
10. **The More Energy to Repair Crew** button brings in more energy for repair processes to work with.
11. **The Prepare Cell for Shutdown** button places the decision to euthanize a cell with its neighbors.
12. **The Master Shutdown** button kills and demolishes the cell.

That's healing at a cellular level

In case you didn't catch it, those buttons are basically healing on a cellular level. Those genetically-controlled processes that the nucleus activates in response to redox distress signals are the mechanisms by which cells are instructed to activate the repair process when repairable, or make way for their replacement when gravely injured.

These are the mechanisms that heal cells and prevent aging. A perfect example is Button #12: The body is supposed to shut down attempts to repair cells after two hours if they were unsuccessful, and then turn on cell suicide (apoptosis). But poor redox signaling fails to make this happen. So damaged cells are allowed to persist and replicate. In a word, that's aging.

Oxidative therapies all operate on the same principles of oxidation and reduction (albeit using quite different approaches)

- Hydrogen peroxide.
- Ozone therapy.
- Hyperbaric oxygen chambers.
- Chlorine dioxide.
- Pulsed electro-magnetic frequency.
- Photonic therapy (infrared light).
- Redox molecules.
- Exercise.

What happens when your redox system fails?

Poor cell to cell communication (referring to both kinds) causes breakdowns and failures to occur in protection, detection, repair, and replacement, which leads directly to

- slow cell repair,
- low energy,
- premature aging,
- fewer species of beneficial bacteria and more pathogens in the microbiome,
- tight junction injury,
- leaky gut and leaky membranes,
- food allergies and intolerances,
- immune system dis-regulation,
- autoimmune problems,
- neurotransmitter and hormonal imbalances,
- psychological disturbances,
- chronic inflammation.

… for real.

This is the very foundation of where all chronic, degenerative diseases come from – lack of communication at the cell level, which is another way of saying *impaired redox functioning.* With poor redox function, cells can't protect themselves from injury. They don't know they've been injured. They don't call for repair mechanisms. And cells don't replace themselves when they ought to.

These imbalances and repercussions express themselves symptomatically wherever the body is weakest or hardest hit by the ensuing hormonal, neurological, and immunological stresses (i.e. rate of injury outstrips rate of repair) – taking their toll as

- poor brain function in autism, Parkinson's and ADD;
- digestive disorders like irritable bowel syndrome and Chron's Disease;
- insulin dysfunction in diabetes;
- nerve impairment in neuropathies; or
- blood vessel problems in heart disease.

All of them are caused by failure of cells to communicate and repair appropriately, which, for all intents and purposes, is the same as *chronic inflammation.* Loss of cell-cell communication is virtually synonymous with chronic inflammation. Chronic inflammation more or less equals loss of communication at the cell level. They're nearly one and the same.

Rate of injury vs. Rate of repair

Cells in your body are constantly being injured and killed every minute of every day. At the same time, the body's healing mechanisms continually repair damage cells and replace dead ones with healthy new cells. So there's always some rate of injury happening versus some rate of repair.

That ratio determines your resistance to dysfunction and disease. It greatly influences your energy level – physical and mentally. It controls your recovery time from exertion, injury, and illness. It has a lot to do with your rate of detoxification. And it's a primary factor in the rate at which you age biologically.

Point being, both of those states – damage and repair – can change quickly through interventions, or gradually through natural processes. But you seldom ever notice any of these changes because the human body has a built-in buffer zone of coping mechanisms – a reservoir of healing capacity – that's designed to favor life and healing until your compensation mechanisms are stretched past their limits of competency.

So you only notice three conditions:

1. When repair has fallen far behind and injury is clearly winning (aka "disease").
2. When your rate of new injury increases rapidly due some event or circumstance (e.g. major infection or massive toxin exposure).
3. When your rate of healing improves quickly from a lifestyle change like a healthy new diet, a detox program, or some therapeutic effort.

You see, all life forms are designed to start their lives with a full bucket of life force, if you will. They spend the majority of their lives with healing capacity to spare in the form of compensation mechanisms, emergency procedures, and various workarounds that the body uses to cope with less than optimal conditions.

It's only when injury exceeds the body's ability to repair by a wide margin that you see acute disease take place, or you die. Nowhere is this teeter-tooter effect better portrayed than the classic image of a bed-ridden, comatose, gravely ill person who's fighting for their life after being given their last rites.

The determining factor in whether they live or die is which of these will ultimately prevail: repair or injury. Not discounting supernatural influences, the phrase "now it's in God's hands" is another way of saying 'injury and repair' are locked in a fight to the death … or life, literally.

In daily life, however, you typically don't even notice gradual augmentation or erosion of your healing or repair because your buffer zone absorbs slow changes with any excess healing capacity you may have. Your body just does its normal healing thing and you're none the wiser.

But our multitude of sins against Nature have stretched our coping mechanisms to the limit

Our irresponsibilities and excesses have exhausted our healing capacity at all levels – cell, organ system, individual, society, and planet. So now the majority of people are living their lives on a razor-fine edge between sickness and health, with dangerously little safety margin to spare.

That's important because the slightest new insult in diet, lifestyle, or internal milieu can push a person over the tipping point into dysfunction and disease … which we're now seeing explode in our lives, families, doctor's offices, public places, the news, and every health statistic you'd care to measure.

To venture a guess, probably 35% of the population is slightly over the edge into the disease/dysfunction side (minor disease), 15% is well into disease (intractable disease), 35% is slightly on the desirable side of health (occasional health problems/annoyances), and 15% of the population has a fair amount of excess healing capacity to overcome further insults.

In other words, almost everyone is on the verge of manifesting a chronic, degenerative disease – they just don't see it that way. So as soon as the body uses up the last trick it has in its arsenal to keep you running incident-free, that's when serious disease suddenly appears out of nowhere – and won't go away.

Chronic inflammation drives all degenerative diseases

Healers of every modality and mindset are slowly, but surely, coming to the realization that all chronic, degenerative diseases are driven by unrelenting inflammation. Chronic inflammation is directly responsible for doing the vast majority of immune system dis-regulation, cell-cell communication failure, mitochondria injury, metabolic disruption, DNA replication error, and cell destruction.

Therefore, the majority of damage that a disease causes is not done by the imbalance, symptoms, or germ itself, but by inflammation that the body employs to fight the disease. In other words, most infectious pathogens, disease processes, and toxins are not what harm cells, structures, and systems the most.

Rather, the body's own threat annihilation system – inflammation – is far stronger at wiping out friendly cells than the menace itself. Inflammation itself is the mechanism that, when allowed to persist (i.e. chronic inflammation), causes the symptoms we collect into categories and call disease.

In fact, the more perceptive practitioners among us have come to realize, barring trauma, most modern diseases – as defined by symptom clusters, rather than root causes – are nothing more than a variety of different locations and ways that chronic inflammation can be exhibited by the body, as repair fails to keep up with injury.

Simply put, failure of cells to repair is what causes all disease. And that's caused by poor redox signaling, which is partners in degeneration with chronic inflammation.

Cancer cells are the quintessential example of what happens when inflammation persists and cell's distress signals go unanswered.

Cancer is believed to take place after a cell has borne the brunt of some 20-25,000 unrepaired injuries to its DNA (i.e. failure to repair/replace).

How inflammation repairs damage

The body uses inflammation to repair pretty much any sort of damage that can happen anywhere in the body – including

- cuts, bruises and overuse injuries;
- damage to bones, muscles, tendons, blood vessels and internal organs;
- infection;
- radiation damage;
- chemical and heavy metal poisoning;
- as well as ordinary wear and tear.

It does this by first increasing permeability of blood vessels so that immune cells like white blood cells can pass through the vessel wall to fight invading pathogens. This also lets more blood into an area, which we call swelling. Furthermore, swelling makes movement painful, thus limiting mobility and further damage.

Inflammation then launches an oxidative attack on everything in the area it doesn't recognize as your own healthy cells. Through oxidation, the first stage of inflammation kills infectious microorganisms, neutralizes toxins and heavy metals, and it destroys your own damaged, diseased and dead cells.

When inflammatory processes are working correctly (i.e. acute inflammation), the first phase of inflammation eventually shuts down and the second phase takes over. In the second phase of inflammation – the repair phase – unhealthy cells and debris are replaced with new cells and collagen fibers (made from protein and cholesterol).

On the other hand, when inflammation doesn't stop (i.e. chronic inflammation) both phases are active at the same time – destruction and creation. And that's the leading cause of chronic, degenerative disease today – inflammation that doesn't stop destroying tissue.

Acute inflammation vs. Chronic inflammation

There are two kinds of inflammation: Acute inflammation, which is temporary and beneficial, and chronic inflammation, which is persistent and harmful.

Acute inflammation heals and protects you on a daily basis from threats that can damage tissue such as cuts, bruises, infectious organisms, ordinary wear and tear, oxygen deprivation, and toxin exposure. Common occurrences like exercise and sun exposure also turn on inflammation without you ever knowing it.

Through acute, "fast-acting" inflammation, the immune system gets called into action. It wipes out the threat in a matter of hours to days using a four-step process whereby (1) oxidants *destroy*, (2) antioxidants team up with reductants to *neutralize* the oxidant, (3) redox signaling partners with the immune system to *repair/replace* damaged cells.

Then (4) when that process is finished – if everything's working properly (i.e. complete healing and the redox signaling to match) – inflammation shuts down, and the body resumes normal operation.

On the other hand, chronic inflammation is where the same processes get activated, but they are unable to turn themselves off. Step 4 *shut down* gets started, but never gets turned off completely. Instead, the first three stages get stuck in a vicious circle of destroying and repairing cells repeatedly, which can smolder along unbeknownst to you for years or decades.

The reason the immune system is not able to turn itself off after the initial burst of activity is one or more of the following circumstances:

#1 – Overwhelmed antioxidant system

First, some significant insult takes place requiring the immune system to use its oxidative stress tool – often between ages 10-30. This could be a sports injury, a major infection, or an acute poisoning event (like injury resulting from a major round of vaccines).

In this situation, cells don't have antioxidant and reductant reservoirs large enough to neutralize the oxidation being done. So the repairing and replacing of damaged cells is done slower and less completely than it should.

Cells then eke by and reproduce themselves in a partially damaged state, which is frustrated by lingering attempts by the immune system to simultaneously *destroy* with low-grade oxidative stress and *heal* using redox signaling and repair processes.

This smolders along below your awareness until, later in life, your cells are being injured by oxidative stress faster than they're able to heal. At that point, you get symptoms that bother you. But all the while you've had chronic inflammation that only bothered you occasionally (like an old knee injury that hurts only when cold weather returns).

#2 – Unending exposure perpetuates inflammation

The insults that triggered the acute inflammation persist for one reason, or several reasons, to complicate matters. Examples would be

- you continue to eat pro-inflammatory foods like fabricated vegetable oils;
- Corrupted Gut causes poor digestion, leaky membranes, food sensitivities, immune system hyper-activation, and autoimmune conditions;

- you continue to stress that old knee injury with the help of painkillers (or another medication that adds fuel to the inflammation fire);
- you come down with a long-term illness like Lyme;
- or you continue to take in toxins faster than you release them.

#3 – Fewer reductants, diminished redox signaling capacity, and shortage of energy prolong inflammation

As we age, we lose mitochondria. Less mitochondrion mean fewer reductants are made, diminished redox signaling capacity, and less energy for all bodily processes. Plus, the mitochondria that do stick around age with us. Their DNA gets damaged just like ours does.

As our mitochondria age, our once-balanced blend of oxidants to reductants tends to tip toward a surplus of oxidants, and a deficiency of reductants. That pushes us increasingly in the direction of oxidative stress as we grow older. (However, too much of either one – oxidation or reduction – results in unregulated oxidation and cell damage.) In addition, inefficient redox signaling sends unclear messages to the nucleus, which then makes fewer antioxidants.

So in this situation, any sort of severe or prolonged damage can cause the immune system to dump oxidant in an area that the cell's not able to clean up quickly due to a deficiency of reductants and/or antioxidants. As oxidative stress, positive charge, and acidification build up in the cell, the clarity of the message conveyed by redox molecules declines.

From that point on, the immune system's oxidative response becomes its own worst enemy. Oxidant is dumped in the area to start the clean-up process, but the cells don't have antioxidant reservoirs large enough to counteract the oxidation effect.

Full healing never takes place. The immune system senses something's wrong, but it gets confused by the mixed signals and failure of its tools to do their jobs properly. So it continues to promote the destructive component of inflammation as it tries in vain to rebuild as best it can.

Neither side – destruction or healing – is able to seize control while the immune system is caught in this vicious cycle of antioxidant/reductant deficiency and imbalanced/unclear redox messages.

That's how acute inflammation turns into chronic inflammation. And that's how chronic inflammation causes disease: It's the failure of cells to repair or replace themselves efficiently and effectively. Thus cells continue to live and reproduce in an unrepaired state.

Most important, that's a large part of what drives all the degenerative diseases you know by name, including heart disease, arthritis, Alzheimer's, cancer, diabetes, and GAPS conditions.

Repeatedly icing painful joints undermines the healing process

Some people ice and wrap swollen areas of their body to reduce swelling. This interrupts inflammation and shortcuts the healing process. For instance, you'll see many NBA players icing their knees when they're taken out late in a game and don't expect to return.

This restricts blood flow to the area in an effort to ease the pain, maintain their mobility the next few days, and reduce their downtime. But using this trick to keep them in the game and productive often comes at a cost, because it defeats the healing benefits that inflammation is trying to bring to the area.

It makes them feel better short-term so they can continue playing more pain-free during the season. But minute damage tends to accumulate over the course of their hectic schedule and turn into weaknesses and susceptibility. And that leads to injury – both acute injury and the repetitive/overuse variety that often returns to haunt them (1) during their career when they least expect it, (2) later in their career, or (3) later in life.

Insulin triggers inflammation, and redox molecules mediate hormone response

The hormonal system (endocrine system) regulates blood-sugar levels by releasing varying amounts of insulin, which unlock the "doors" that drive sugar into cells. Insulin also causes the liver to store sugar for future use.

When you eat lots of sugar, mitochondria crank out redox molecules as a by-product of ATP production. Problem is, the more corrupted your energy production system gets to be, the more oxidant tends to be made relative to reductant.

That extra oxidant increases the build-up of uncontrolled oxidative stress, which can threaten cells all over the body with oxidative damage, including insulin-producing beta cells in the pancreas.

To protect itself from this rapid burst of oxidative damage, the pancreas is programmed to pump out insulin as fast as it can in order to store that sugar in the liver, instead of burning it immediately and overwhelming the system with oxidants.

Unfortunately, it has the opposite effect. What ends up happening is insulin drives the sugar into cells faster and more furiously, which only increases oxidative stress and inflammation. So basically, eating lots of refined carbs and sugar is similar in a lot of ways to pouring gasoline on a fire. It's a sugar-induced explosion of oxidative stress that can only be controlled (hopefully) by flooding the system with insulin.

This isn't a problem when antioxidant supplies are capable of handling the increased load. But, more and more often, that's not the case as our population loses its healing capacity (yet another coping mechanism stretched to the limit). Thus you've got yet another factor, among many, that increases inflammation.

This is one of several vicious cycles involving redox reactions that elevate and perpetuate inflammation. In this case, with antioxidant reserves exhausted throughout the body, insulin conspires to keep the destruction phase of inflammation going longer than it's supposed to.

In short, frequent high levels of insulin cause the battle between inflammatory *destruction* and *repair* to rage on continuously. Of course, excess insulin also leads to diabetes (another pathway by which poor redox signaling turns into disease).

Disease is made possible by a shortage of cellular energy

Lack of energy at a cell level is involved in the inception and continuation of virtually every disease process. When cells don't get enough energy to function properly, they malfunction. They then react, interact, and replicate improperly. This is one of the root causes of dysfunction and disease, among several interrelated conditions.

That is, cells consume ATP for energy. ATP is made in the cell by mitochondria. So mitochondria are the power plants of the cell. Unfortunately, a large percentage of GAPS individuals are genetically predisposed to having mitochondrial insufficiency.

In other words, they can't make enough energy to satisfy all the demand. And when a resource is scarce, something's got to give. In autism, that often means the body will then rob from "brain function and detoxing" in order to allocate whatever energy it has to "basic life support and gratuitous muscle activity."

To complicate matters further, cells that can't make enough ATP revert to a much less efficient fallback mechanism of fermentation (called glycolysis), which is sugar-based energy production. Energy-starved cells basically resort to consuming sugar in an effort to sustain themselves. And that manifests as sugar and carb cravings, which contributes to gluten sensitivity and candida overgrowth.

Largest energy-consuming processes in the body

If I had to guess, this is how I imagine the body normally allocates its energy:

1. **Basic life-support.** Heart, blood vessels, lungs.
2. **Physical activity.** Muscles and nerves.
3. **Brain function** (5% of body weight, but consumes 20% of the body's glucose).
4. **Digestion.**
5. **Production/rebalancing of biochemicals.** Neurotransmitters and hormones.
6. **Cell replication.**
7. **Fighting infection.** Innate and adaptive immune response.
8. **Muscle repair after exercise.**
9. **Fighting aging.**
10. **Detoxing.** Heavy metals, chemicals, metabolic waste products.

Of course, this is a highly dynamic list that's always fluctuating. The body definitely changes its priorities in response to shifting demands, which makes for occasional conflicts, postponements, and cancellations of bodily processes.

For example, the tummy will put digestion on hold when you exercise intensely on a full stomach. Mental acuity drops in the latter stages of an endurance race when you need that energy to power muscles. And, the granddaddy of them all, the body virtually throws this list on the back burner and activates survival mode when it's fighting a major illness like cancer or the flu.

Point is, quite a few systems and processes are vying to get their piece of the limited supply of energy the body is able to extract from food, air, water, etc. And when there isn't enough to go around, the body has to make compromises. It has to make choices and accept sacrifices. This comes into play in a big way when a major health challenge chronically restricts energy production throughout the body

The best example occurs in autism. The factors that combine to cause autism invariably lower energy production throughout the entire body, and increase demand, if anything. You can tell how bad the discrepancy is in autism by how many areas are hit by shortages, and how high up the ladder they go. Nearly every physical and mental process is compromised by lack of energy … and it shows.

What are mitochondria?

Mitochondria are the power plants of cells. They are essential to the life of every multi-celled organism (e.g. humans, animals, plants, protozoa) because they make the only form of energy their host's cells can use – ATP. Cells, mitochondria, and the body also talk with each other via molecules made during ATP production, called redox molecules.

About 200 to 5000 mitochondria live inside most of your cells. They take care of the cell in so many ways (or not). However, they are their own life form (organelles, actually) and possess their own DNA. So they ride along inside each cell, yet they reproduce, live, and die separately from the cells that host them.

Somewhere along the line, eukaryotes (multi-celled organisms) and mitochondria made a deal to work together in a cooperative arrangement. Mitochondria agreed to take up residence inside the cells of eukaryotes, get fed on a regular basis by the organism, and, in exchange, they produce ATP to power our cell's activities.

Basically, the more mitochondria you have, the more energy you can produce, and the more efficiently your cells operate in so many ways. Conversely, depressed mitochondria populations equates to chronic low energy (like chronic fatigue syndrome), poor cell-cell communication, poor cell repair, and accelerated aging.

Unfortunately, mitochondria populations naturally diminish (1) as you age, (2) when their communication network breaks down, (3) when toxins upset them, (4) when the both of you are malnourished, and many other things.

How mitochondria make redox molecules

Animals like us can't access any energy directly from the food we eat. Instead, we need mitochondria, and a number of processes, to make our fuel for us (called ATP).

First, bacteria and your digestive system have to break down the food you eat. Then, the liver converts those food constituents into two things – fat or sugar. After finding their way into cells (which can be a challenge when you've got insulin problems), mitochondria burn that fat or sugar in the presence of oxygen and turn it into electrons.

As electrons are moved through the electron transport chain to make ATP, the process produces redox molecules as a by-product. Note how many steps there are between the food you eat and fuel your cells can actually use.

Unfortunately, as we age, we lose about 1% of our mitochondria population per year. And with their demise, we lose their energy production and redox signaling capacity. So by the time we're 70, we're down to about 10% of the energy output, and healing capacity, that we had as teenagers.

And since mitochondrial redox molecules are the communication network of cell repair, at least 50% of the aging process is a loss of redox molecules.

Both cell-cell communication networks come from mom

Curiously enough, you inherit all your mitochondria from your mother. They're present only in the egg. There are no mitochondria in sperm. Mitochondria then reproduce as welcome guests inside each new human cell made. So you acquire half your cellular communication ability from your mother in the form of mitochondria and their oxygen redox molecules.

You inherit the other half of your cell-cell communication ability from your mother in the form of bacteria and the carbon redox molecules they make. As described in the chapter "How the Microbiome Gets Established, and How It Gets Corrupted," the microbial starter cultures that seed your microbiome come from mother's birth canal as you're being born (natural childbirth).

Those first bacteria populate the GI tract and make the carbon redox molecules that form the communication network of the microbiome. So as it turns out, fathers are almost useless in the world of communication… on a redox level. Their role happens very much on the periphery of all the action.

The human body has incredible machinery to repair itself, but it's useless without effective redox signaling

Our DNA is the same as it was decades ago. Our machinery for cell repair is the same. Our enzymes for detoxing haven't changed. Everything is the same as it always has been. But they're not doing their job.

It's all sitting there ready to kick into action. But if the signaling network goes down, cells can't protect themselves like they're supposed to. Perhaps even worse, cells don't even realize when they've been injured, let alone call for help.

With our rate of injury so high, and our rate of repair so low, cells are not repairing and replacing themselves like they used to. Our cells are now accumulating toxins and repeated damage to DNA, rather than constantly rejuvenating themselves, as they're designed to do. So those imperfect cells persist and replicate in an unrepaired state, which we call aging.

The smartest scientists in the field said it couldn't be done: Stable redox molecules outside the body

Thousands of researchers have been studying redox reactions in earnest for the past 10+ years. There are now some 100 scientific papers being published in peer-reviewed journals each month on the subject of redox reactions, representing about 300 months' worth of research coming out every 30 days.

And the prevailing opinion was that it couldn't be done: Redox molecules stabilized outside the body. Reason is, when mitochondria make reactive oxygen species, they only live a fraction of a second. They're so reactive they bond immediately and they're gone. So the idea of stable redox molecules, in any way, shape, or form, was preposterous to the researchers who would know.

But that didn't keep one adventuresome startup company from trying: Unfortunately, after 12 years and $15 million dollars of research, the best they could do was keep oxygen redox molecules stable for ten minutes. But that wasn't long enough to make a viable product people could benefit from. They went bankrupt.

With a good head start, Gary Samuelson, PhD (medical atomic physicist) did the impossible: Having access to the first team's research materials, Dr. Samuelson took 18 months to come up with a method to keep oxygen redox molecules stable for well over a year.

At a half-life of two years, Samuelson created a viable product that could actually help people. The product, called Asea™, was launched in 2010 as a network marketing company. It was the first redox signaling molecule supplement of its kind.

Asea (for mitochondrial communication)

Redox signaling molecules, in supplement form, offer humankind a monumental breakthrough in healing, cell-cell signaling, anti-aging, and human health in general, because redox signaling is so fundamental to everything that cells do.

To a large degree, the redox signaling system can either giveth life when it works well, or it can taketh away when it fails to perform as well as it did when we were young.

Asea helps

- **Protect.** Increase antioxidant effectiveness by 5X, thus raising your defenses against infections and toxins.
- **Detect.** Make cells aware when they're damaged.
- **Repair.** Activate genetic switches to make new proteins to repair damaged cells.
- **Replace.** Prompt un-repairable cells to kill themselves, and tell neighboring cells to fill in the void.
- **Hormone reception.** Normalize response to hormones.
- **Detoxification.** Up-regulate detox pathways through NRF2 activation.
- **Energy.** Oxygen redox molecules regulate mitochondrial population and metabolism, thereby increasing energy production in the cell.

Independent scientific research conclusively demonstrates Asea increases athletic performance (i.e. VO2 max, ventricular threshold). One mechanism is by mobilizing fatty acid stores (presumably from the abdomen) for cells to burn during exercise (which is about 20 times more efficient than burning glucose).

Simply put, Asea can take years to decades off your "cellular age" (i.e. how well your cells work) in a middle-aged to senior adult. **Incredible but true, in one week on Asea, you can see a 10-30% increase in mitochondrial population.**

That means you can reverse roughly 10-30 years of aging in terms of mitochondrial signaling capacity (e.g. healing), energy production, and much more. Now that's a profound supplement. And that's why many people – academics and users alike – consider it to be miraculous.

Though children and teenagers whose redox signaling systems still work well probably won't notice a dramatic difference.

Diversity of species and fluent communication gives the microbiome its strength and resiliency, not the absence of pathogens

So this is one of the rare cases where the alternative/holistic crowd, and the conventional medicine crowd, are both going to discover they were close to the truth about the microbiome. However, they were both off-base just a little bit with their beliefs and therapies, because they lacked crucial pieces of information.

A few years from now, people are going to realize all practitioners in the 90's and early 2000's made a number of incorrect assumptions about why gut dysbiosis compromises your health and indicated certain treatments.

To give you some idea what happens when you're in the right ballpark, but you're not quite hitting the nail on the head, let's look at a few of the popular approaches practitioners have used to improve digestive health in the decades leading up to said discoveries... and how they've worked out.

The long-term benefits of digestive remedies has been disappointing (e.g. probiotic supplements, acid blockers, acid supplement (HCL), and digestive enzymes)

When probiotic supplements first became popular in the 1990's and 2000's, clinicians thought they'd do a lot more good for people than they actually have. Eager to use this promising "new" healing tool, practitioners of all types recommended that their clients take probiotics to rebalance their gut flora and reverse their chronic conditions.

But the results have been mixed. Many would say their pathogen overgrowth like candida seemed to recede while taking the probiotic. Many would say their symptoms subsided and they felt better for a while, but not necessarily well. Clinically, a modest percentage of people in the thousands of studies showed marginal results.

Supplements may have helped a good number of people, a decent amount, for a reasonable amount of time. But most would say, probiotic supplements don't get your gut anywhere near fully healthy as a standalone or complementary treatment.

They haven't met people's admittedly high expectations when the theories were first raised in public. Probiotics weren't the complete solution people were looking for. And the probiotic brand/formulation they took is partly to blame for the disappointing results.

That is, impotent probiotics (for whatever reason) can only give you limited benefits. But probiotics made from wild sources will naturally have wider diversity (which, like all naturally-sourced supplements, can't be quantified on the label the way synthetic/processed supplements can). Wild-sourced probiotics can get you much closer to the results you would hope for.

Digestive enzymes, HCL, and acid-blocking drugs have shown similar results

They've shown decent utility among many folks. But none of these has earned a reputation as a go-to solution for flawless digestive health. Reason is, probiotic supplements, acid blockers, acid enhancers, and digestive enzymes are attempts to solve the wrong problem. They're close. And you may get decent results being in the right ballpark.

But restoring cell-cell communication looks like it's more valuable to the microbiome than simply driving pathogens out, or covering up symptoms for a while. Diversity appears to be what the body's looking for, even over strength of individual bacterial strains.

In contrast, almost all commercially-produced probiotic supplements are made from a very limited number of species of bacteria (usually 2-7, sometimes as much as 24) originally taken from cow intestines and cultured in isolation. Problem is, those bacteria are not prevalent colonizers of the human gut due to vast differences in pH, digestive enzymes, and overall design (four stomachs instead of one).

But as crazy as that sounds to be ingesting cow bacteria, they can provide some benefit for a really screwed up microbiome that's filled with weeds of the gut (e.g. candida, klebsiella, pseudomonas, E. coli). They can do some of the things our native bacteria do. But nowhere near the whole shebang of systemic benefit we rely on to have complete health.

Bottom line: It's better to get the biodiversity back in the microbiome than the artificial environment of (most) probiotic supplements, digestive enzymes, acid enhancers, or acid blockers.

Wild ferments are the exception to the rule

Most store-bought fermented foods such as yogurt, sauerkraut, and kimchee are made from a very limited number of bacterial species added as starter culture (usually one species). This applies pressure to the microbiome to narrow its diversity of species.

Yogurt's the naughtiest one here due to its very limited number of bacteria species (usually one species), high sugar content, casein, and bare minimum fermenting time. Plus, yogurt's digestive benefits tend to be greatly oversold by food companies, giving you a false sense of security.

In contrast, you don't add any starter culture to make a batch of wild fermented vegetable. You simply let the microbes that found their way on to the surface of the vegetable do the work. So you may get hundreds of potent, soil-based microbes to join the party.

Bottom line: Wild fermented veggies – great for you. Standard fermented veggie products – good for you. Conventional yogurts and fermented dairy products: generally neutral overall to detrimental.

Bacteria use carbon redox molecules as both a communication network and a nutrient-rich soil

Bacteria need to communicate with each other, the gut lining, and surrounding organ systems (like the gut associated lymphoid tissue, or GALT), to let their needs be known… almost like they function as one organism (hint, hint).

But bacteria don't have mitochondria. So they make their own communication network comprised of redox molecules with a carbon backbone, instead of mitochondria's oxygen-based redox molecules. Produced by bacteria's metabolism, these carbon redox molecules send and receive information related to the wellness or sickness of various cells in the microbiome.

All signaling between microorganisms happens through carbon redox molecules, thus the awesome intelligence of the microbiome is transmitted through carbon redox molecules. Rich in minerals, carbon redox molecules also provide the nutrients that feed bacterial growth.

So bacteria take care of the health of the microbiome, the gut, and the entire organism through their communication network of carbon redox molecules. Carbon redox molecules are the microbiome's greatest strength when diverse and fully fluent. Conversely, they're the microbiome's glaring weakness when depleted and uncommunicative.

How bacteria make and use carbon redox molecules

Each of the tens of thousands of bacterial species makes its own family of about 10-15 distinct carbon redox molecules, resulting in hundreds of thousands to a million unique variants of these carbon redox metabolites.

Dubbed "carbon snowflakes" due to their extremely varied molecule structure reminiscent of conventional snowflakes, carbon redox molecules are both a communication medium for cell-cell signaling, like a smoke signal. And they're a source of minerals for microbiota to grow in, like compost-rich soil for gut flora.

Each type of carbon snowflake has its own specialized role to play in the language of cell-cell communication. Like words of a language, each variant may communicate a slightly different message, be used for slightly different purposes, and do things others can't.

That fluency permits an unimaginable variety of messages to be transmitted and interpreted loud and clear. For instance, rare species of bacteria, possessing their own unique abilities, may communicate using their own family of scarce redox molecules.

When those redox messengers are present, depleted populations realize they're endangered and can then send out a signal to neighboring species requesting space in the microbiome to grow. Other species give them that room, and they go into rapid replication mode.

But if that redox smoke trail is absent, those bacteria (and the group) can't tell how many others of their species are living in the gut (it's a big place to them). And without any way of talking amongst their family, they're not able to sense when their kind are depleted and need to repopulate. Hence, they stay an endangered species, or go extinct. That's one of the main mechanisms by which diversity in the gut has plummeted.

On the other hand, when probiotic bacteria sense an overgrowth of pathogenic microbiota like candida, they'll sound the alarm for the whole probiotic team to start making more anti-microbial compounds (anti-fungal, anti-viral, anti-bacterial, etc.) to knock the bad guys back.

Those are two of the most important functions of carbon redox molecules – supporting diversity of beneficial bacteria, and suppressing pathogenic microbes in the microbiome.

However, when you lose biodiversity in the microbiome, you lose that fluent communication our microbiomes once had many decades ago. Like a cell phone with poor reception, bacteria then can't talk to each other and figure out what they or the group need.

Bacterial redox molecules are also the nutrient medium from which microbiota receive minerals to grow. Like compost of the gut, carbon redox molecules are the bacteria's broken down food particles that feed biology in the gut ecosystem.

The human digestive tract used to host about 20-30,000 species of bacteria

This booming ecosystem in a healthy person once contained the biodiversity you might see in a coral reef or thriving rain forest. However, all the questionable choices we've instituted as a society, and as consumers, have taken a huge toll on that variety and balance.

Assailants include

- chemical fertilizers,
- herbicides like glyphosate,
- antibiotics in meat and medicine,
- pharmaceuticals like NSAID's, birth control pills, and steroids,
- heavy metals and other toxins in personal care products, and
- processed foods.

So today you'd be fortunate to have 10,000 species. This was demonstrated by a 2010 NIH study to map the genome of the human microbiome. It took 44 participants to reach 10,000 species of bacteria.

That narrowing of species has put an enormous dent in the microbiome's ability to protect us from tight junction injury and toxin exposure, as well as take care of the gut, which then takes care of the rest of the body.

Lack of diversity in the microbiome starts with lack of diversity in the soil

For millennia, traditional farming techniques supported diversity in the soil through natural fertilizers (e.g. manure), polyculture crops, crop rotation, and composting. But nowadays, Big Ag is fond of growing one type of plant on a plot of land over and over, using inorganic chemical fertilizers, applying crop amendments like glyphosate, in an increasingly lifeless growing medium – and worse. All this destroys microlife in the soil.

That broken ecosystem in soil then gives the plant fewer nutrients and more toxins for us to ultimately receive. Fewer nutrients in plants mean fewer nutrients for our beneficial bacteria to feed on. That means the rare bacteria which are skilled at digesting rare trace minerals for us – they get malnourished and neglected.

So with nothing to eat – (1), handicapped awareness they're endangered – (2), and us ingesting fewer microbes because of our germaphobic culture – (3), rare species of bacteria are under a lot of pressure to disappear from our microbiomes.

Meanwhile, animal antibiotics and human antibiotics positively devastate our microbiomes directly. Then all the other contaminants in our food, air, water, medicine, and personal care products annihilate all the less resistant microbiota.

Beneficial bacteria are your front line of defense against tight junction injury

Your good bacteria protect you from tight junction damage by gobbling up toxic peptides, and other toxins, before they've had a chance to make zonulin. They control pathogen populations that further stimulate zonulin production (like candida). And they increase production of the DDP4 enzyme that breaks down zonulin after it's been made.

Good bacteria nourish and protect the enterocytes that line the gut wall. They reduce inflammation and immune activation in the GALT. Their redox molecules provide the minerals that help rare species of probiotics proliferate. And, let's not forget, diverse species of bacteria create the carbon redox molecules that are the communication network of awareness, defense, repair, and replacement amongst microbes in the gut.

Every protective membrane in the body is tied together by tight junctions (e.g. gut, brain, blood vessels, kidneys, liver, and more)

Membranes. Membranes are a specialized layer of tissue just one cell thick that acts like a firewall between the outside world (pre-absorption) and you (post-absorption). At just 50 microns, or about half the thickness of a human hair, they fulfill barrier functions against toxins, as well as nutrition, hydration, and detoxification roles, by actively keeping harmful things out, while letting beneficial things through.

Every membrane in your GI tract is protected by these dynamic, intelligent membranes that run from your sinuses and mouth, to your throat and stomach, through the small and large intestine, all the way out your rectum. In fact, virtually every major organ in your body relies on a membrane to separate its innards from its "outards."

Tight junctions. Just as important, every cell in every membrane is held to its neighbor by a set of interlocking protein fibers called a tight junction. These cross-linking fibers remind one of Velcro, because they have the ability to open and close whenever the immune system needs to send its forces through the membrane to fight off a threat on the other side.

Tight junction intelligence. That ability to open and close on purpose is what gives membranes the intelligence it then loses when tight junctions are damaged by toxins that stimulate zonulin production.

Cells communicate through gap junctions. Gap junctions live alongside, and are held together by, tight junctions. Like fiber optic cables connecting neighboring cells, gap junctions are wiring through which cells talk to each other. That means breaking tight junctions "cuts the cord" between cells and organs that need to talk to each other about inflammation, health status, stressors like toxins, nutritional status, etc.

Zonulin is the naughty substance that dissolves tight junctions when it's allowed to run amok. The gut wall makes excessive amounts of it in response to the presence of glyphosate, gluten, cholera, clostridium difficile, and/or candida.

More than any other toxin, it's directly responsible for billions of people now suffering from leaky gut, leaky membranes, chronic inflammation, GAPS conditions, and chronic disease. Sad thing is, probably 95% don't know it.

Sources of tight junction damage

The biggest threats to tight junctions and human health today (by potency and prevalence) are glyphosate (active ingredient in pudnuoR), gluten (its breakdown product is gliadin), NSAID pain-killers like ibuprofen, pathogens like candida, and steroids. Tragically, they're everywhere.

Take the most damaging of them all – glyphosate. It's the active ingredient found in the vast majority of herbicides in use commercially and residentially throughout the world. So it's in our soil. It's in our drinking water. And it's in organic food. It's in our blood. It's in our breast milk. And it's in our urine. You can't avoid it. In fact, recent reports say 75% of rain falling on the planet is tainted with harmful levels of glyphosate.

And we have the US Environmental Protection Agency to thank for setting up the rules and regulations that permit (condone) that reckless assault on the health of every man, woman, and child presently being exposed. It's a horror show of big corporations and government agencies in cahoots we each other.

One instance: Already toxic beyond question, in 2013 alone, the US EPA passed a 30-fold increase in allowable limits of glyphosate on many food crops in the US – all because glyphosate does wonders for the bottom line of those involved. More incriminating evidence to come.

Dr. Zach Bush on *how you can tell if you've got a leaky gut*

> *"If you've got a gut, and you live on planet earth,* **you've got a leaky gut**. *That's pretty much where we're at right now."How glyphosate and gluten cause leaky membranes, chronic inflammation, and degenerative diseases*

How glyphosate and gluten cause leaky membranes, chronic inflammation, and degenerative diseases

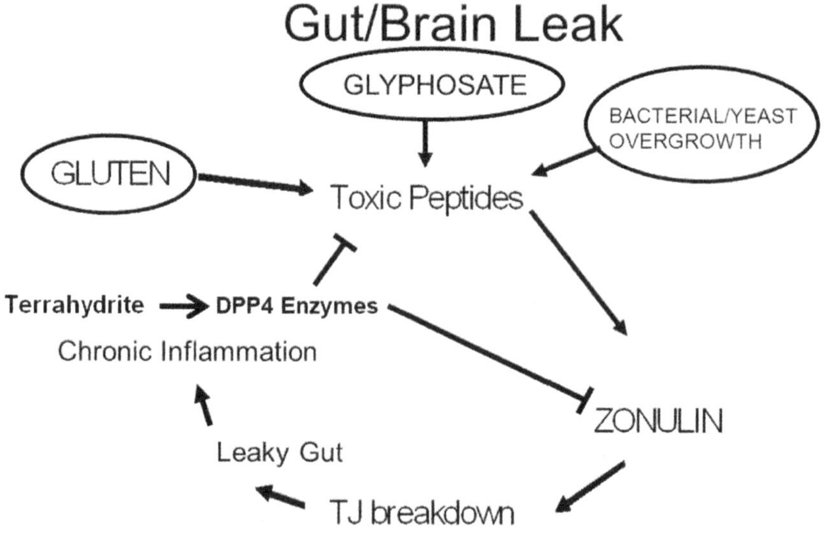

(Graphic used courtesy of Dr. Zach Bush)

The diagram above illustrates the pathway through which toxins in the diet, and pathogens in the gut, trigger a cascade of events that breakdown tight junctions, turn membranes leaky, induce chronic inflammation, and cause dis-regulation and disease throughout the body.

First, gluten (gliadin), glyphosate, and pathogens like candida introduce toxic peptides that make the gut mucosa produce large amounts of zonulin. This zonulin dissolves tight junctions in the gut wall that would normally keep it virtually impenetrable to inappropriate material.

That's what causes leaky gut. That leaky gut then leaks zonulin into the bloodstream, which goes everywhere. Zonulin proceeds to attack membranes all over the body, resulting in leaky blood-brain barrier, leaky blood vessels, kidney tubules, liver cells, and more.

Meanwhile, just behind the gut wall sits the gut associated lymphatic tissue (G.A.L.T. for short) which makes 80% of the immune system's antibodies – which is why they say some 80% of your immune system is located in the gut. Broken tight junctions expose the GALT to every inappropriate substance that enters the digestive tract.

This irritates the lymphatic system, which exhausts its antioxidant reservoirs, dis-regulates the immune system, and unleashes chronic inflammation. And, as just discussed, unrelenting inflammation drives all degenerative diseases, most notably ADD and GAPS conditions.

Immune system irritation in the nose creates seasonal allergies

When you have intact membranes in the sinus passageways, you can breathe pollen and pet dander all day long and not have any problems. But when broken tight junction cause permeable membranes to let allergens into sinus tissue inappropriately, the immune system overreacts.

Responding to the intrusion, the immune system in the sinuses releases antibodies to fight the foreign matter, which triggers the release of histamine. That's what we call seasonal allergies. Of course, conventional medicine combats this effect by blocking histamine, which is the symptom, not the source.

Dr. Bush's new product works wonders in the nasal sinuses and beyond: The Restore4Life team recently released Restore™ in a nasal spray form. It's been flying off the shelves come springtime because it gives immediate relief (within minutes) to those who suffer congestion and post-nasal drip from seasonal allergies.

What's more, it begins protecting and correcting a dysfunctional microbiome at the front line of the immune system's front line – the nasal sinuses – thereby taking a huge load off of your defense systems downstream from the nose (particularly around the stomach, small intestine, and GALT).

Other devastating effects of tight junction injury and leaky membranes

A core component of Corrupted Gut, *tight junction injury* spreads toxicity, imbalance and dysfunction throughout the body. This shows up as a loss of integrity/intelligence to membranes, dehydration, nutrient deficiency, and loss of cellular communication.

Collecting and directing fluids where they need to go: The second way leaky membranes harm cells is by dehydration. When membranes lose their selectivity over what they let through vs. what they block – when membranes lose their intelligence – they lose their ability to control where the fluid you drink goes in the body.

You could be drinking your recommended 8 glasses of water a-day. But if the colon can't pull water out of the food you eat properly, distribute it to cells via the bloodstream, and get it into cells the way it should – all tight junction dependent – you could suffer dehydration in the cells. At the same time, water is the vehicle by which nutrients are taken into cells. So **lower water volume also restricts the flow of nutrients in, and toxins out.**

Mitochondrial insufficiency lowers electrical charge: To make dehydration worse, when redox molecule production in the mitochondria slows down, the cell loses electrical charge. Electrical charge is what pulls water into cell. So as mitochondrial activity goes down, dehydration goes up inside the cell (where it counts).

Nutrient overload and nutrient starvation: Another way tight junction injury harms cells is by nutrient deficiencies and toxic overload. Every type of cell has different nutritional needs.

For instance, muscles may need more proteins (as amino acids). Nervous system cells love fatty acids and cholesterol. Thyroid cells are big fans of iodine. And liver cells can use all the antioxidant vitamins they can get. Point being, cells are picky about what they like and what they don't like.

As a result, leaky membranes cause a surplus of nutrients for some cells that can become toxic to them in excess, while others starve downstream because they were supposed to get those nutrients delivered to them. So no one's happy when membranes fall apart.

Increased EMF sensitivity: A growing segment of the population is becoming sensitive to EMF pollution because of the damage that glyphosate, gluten, and various pharmaceuticals do to tight junctions and gap junctions.

What's happening at a cell level is, glyphosate, gluten, and various environmental exposures break down the tight junctions that hold neighboring cells to one another. When tight junctions disintegrate, the gap junctions behind them that act like fiber optic cables connecting cells – they break apart as well.

So when communication between cells goes down, cells can't synch their vibrational frequency with each other, and the whole fabric of the membrane loses its vibrational coherence. When that happens, stray electromagnetic frequencies become the predominate vibrational signal they're exposed to. So instead of health-giving vibrations supporting a cell's well-being, they're being violently shaken at rates that injure, confuse, and kill them.

Fortunately, Nature stored the antidote to man's own self-destruction in the ground beneath our feet

In 2012, William Vitalis showed Dr. Bush a white paper about soil science in the hopes they could find out why roughly half of Dr. Bush's patients responded well to an aggressive food-based diet, while the other half didn't get better, or even got worse.

As Dr. Bush flipped through the pages, a molecule jumped out at him that looked like the compounds he had been researching involving mitochondrial communication – only this one was carbon-based, rather than oxygen-based. The molecule looked like it had greater redox signaling potential. And it looked like it was significantly more stable than the molecules he had been researching.

Might this have been used for cell-cell signaling by whatever made it? Might it be useful therapeutically if they could find out how to make it and keep it stable? Long story short, it took his group two weeks to find out the molecule found in soil was indeed made by a bacteria … and it definitely had redox potential.

This led Dr. Bush to the astonishing conclusion that bacteria make their own communication network with by-products of their own metabolism. Namely, they make carbon-based redox molecules when they digest their food.

Now a big piece to the whole picture is bacteria are prokaryotes. That means they don't have mitochondria. So they can't communicate using mitochondrial (oxygen-based) redox molecules. However, bacteria still need a way to communicate with each other and their host's cells.

So both in soil, and inside you, they use their own metabolic end products to form their own communication network, which are carbon redox molecules. That's how they talk to each other in the same way that mitochondria send and receive messages using their metabolites – oxygen redox molecules.

In other words, the molecule depicted in the white paper led Dr. Bush to the stunning revelation that bacteria have a communication network, first of all. And it's the same whether the bacteria live in soil or in a microbiome (whatever creature that may be).

This was a brand new idea to medical science. The field was already studying mitochondria and their redox signaling molecules into oblivion. But bacterial redox molecules had yet to be discovered and studied as such. That begged the investigation into what carbon redox molecules are designed to do, and what they potentially could do when put to good use.

Dr. Bush's team invented a way to make carbon redox molecules

Turns out, each bacteria species makes its own family of these carbon redox molecules. So when you have an ecosystem as diverse as that found 200 years ago, before industrialization, there might be upwards of a million variants of these signaling molecules.

Knowing it would be darn near impossible to find someone today with a perfect microbiome, Dr. Bush set off on a quest to find soil that had the diversity he was looking for, yet hadn't been tainted by modern pollution.

That ruled out wet climates, and all soil exposed to today's atmosphere. So he ended up finding sediment buried under the desert southwest of the United States to find the highly compressed, now fossilized remains of soil. Dating back 50 million years, this layer of "lignite extract" is 8 feet thick and biologically inert when it's pulled from the ground.

He took that carbon-rich sediment to back his lab, and a colleague of his taught him how to reanimate its bioactivity by restoring the mineral ratios and amino acid balance the soil once had back to that of living soil.

When he did that, voila, the oxygen and hydrogen binding sites the carbon molecules use to communicate with regain their biological activity. What's more, the compound stays stable. So finally, here was the very same signaling system bacteria use to communicate with each other – in a bottle, and shelf stable. Think of what it could do.

Redox communication in action

First order of business with any new substance intended to heal people is to test it for toxicity. So Dr. Bush put the compound through the industry standard test.

He applied the compound to kidney tubule cells and was amazed (but not really surprised) to find out it's one of the least toxic substances ever discovered, considering how biologically active it is (i.e. potency). It's far less toxic than water. You can drown in it, but exposure to ridiculously high concentrations just makes extremely sensitive cells live longer.

Next thing he did after getting carbon redox molecules stable in a bottle is he went straight to his lab to see what the compounds did to cancer cells, because that's his background. At this point, he wasn't thinking about what this discovery might mean to the microbiome and systems that depend on it.

After exposing cancer cells in a petri dish to these stabilized carbon redox molecules, Dr. Bush saw a burst of reactive oxygen species shooting out from the mitochondria, which means the need for more oxidation was detected and deployed. Then he put the compound on healthy control cells and the exact opposite happened: their ROS production went down, which means less oxidation was needed and produced.

In other words, if cells were healthy, the immune system turned down the stress load placed on them by oxidation, which is destructive. But if cells seemed to be hurting, more potent healing mechanisms were deployed to accelerate repair and replacement of cells (i.e. turned destruction up).

This shocked and amazed him and his team even more than the toxicity test, because it was the first time anyone had seen one supplement produce two completely different reactions based on how stressed the cells were at the moment, and what they actually needed.

He points out Restore™ does not treat cancer, or any disease for that matter. Instead, it re-establishes the redox signaling network by which bacteria communicate, proliferate, and support the health of the gut. And that helps the body heal itself.

Dr. Bush makes a connection with monumental implications

The carbon redox molecules he and his colleagues were able to stabilize in the lab were the very same molecules that bacteria use to communicate with each other and our own cells. Mitochondria do this with similar molecules they produce. So bacteria must be doing the very same thing with their carbon metabolites.

Further study has revealed this is indeed the case: Bacteria use these carbon molecules for cell-cell signaling. And the diversity found in ancient ecosystems can be supplied as a supplement. Keep in mind, in 2012, the concept of stable, renewable carbon redox molecule were brand-new to the literature.

Eventually the magnitude of the discovery hit him: This could have monumental implications to the health of people around the world, on every level imaginable – individual microbes, microbiome, systemically, family, healthcare system, society, and planetary population.

Dr. Bush then asked a seminal question:

> *"What happens if we stop trying to micro-manage and over-engineer our gut bacteria (with probiotics, etc.), and instead just give back the cell-cell signaling the gut is supposed to have?"*

Answer: Put the communication network back in and immediately everyone knows what the other one needs. Restoring the redox signal communication network dramatically increases the body's fighting, coping, and recovery mechanisms. The compound literally turns the clock back many years to decades on your body's ability to heal and protect itself from injury.

The carbon redox molecule supplement now called "Restore™" (generically called "terrahydrite") completely blocks tight junction damage all over the body. Toxins are blocked from getting in. They're neutralized in the gut probably 10 to 50 times better. Toxins can be collected in detox organs and eliminated properly. And the strength of the microbiome is revived. All it takes is fluent communication to make it all work again.

Restore™ (for bacterial communication and gut health) does at least six things to repair tight junctions and create a thriving ecosystem in the gut

1. Breaks down toxic peptides that trigger zonulin production.
2. Breaks down zonulin itself.
3. Restores proper balance of minerals in the microbiome so enzymes can work.
4. Restores the communication network between bacterial species.
5. Increases glutathione production 1200% in the gut.
6. Increase stem cell activity by 30% in the gut.

The carbon redox molecules in Restore™ enable the thousands of species of bacteria to talk to each other and the immune system. That helps create balance and diversity in the microbiome, which creates a cascade of positive protective and restorative effects to happen in previously corrupted systems.

Fluent communication between bacteria lets them first become aware when beneficial species are depleted (particularly rare species). This makes them go into rapid replication mode, bringing balance back to a previously Corrupted Gut. For instance, fluent communication system makes the group aware when pathogenic microbiota have overgrown, so they can produce more anti-microbial compounds to drive pathogen populations down.

Restore™ also rebalances the mineral composition of the microbiome to that of a healthy gut – somewhat like a bio-friendly fertilizer for the microbiome. This makes micronutrients available for the production of enzymes that drive most cell activities.

Finally, Restore™ increases breakdown of toxic peptides that stimulate zonulin production – before zonulin is ever made. And it breaks down zonulin after the gut lining makes it. These actions and reactions give the microbiome the resources it needs to overcome communication blocks, a broken ecosystem, and tight junction injury.

The microbiome can then revive its balance and numbers from the 5-10,000 species it now has, to something closer to 20-30,000 species people used to have. Our microbiomes then have the tools to prevent or repair injury from the most damaging tight junction toxins known to man – glyphosate, gluten (gliadin), NSAID pain killers, and steroids.

Two more mechanisms by which terrahydrite cleanses and repairs: When terrahydrite touches the wall of the small bowel, there's a 1200% increase in glutathione production in a matter of hours, which is the body's most powerful antioxidant. Last but not least, one tablespoon of terrahydrite increases stem cells activity by 30%.

Bottom line: When the gut has what it needs to thrive – that is, when man gets out his own way and lets Nature do what it's designed to do – the body's innate intelligence knows what needs to be done, and it has the machinery to do it. Restore™ helps do that while toxins are an unavoidable part of our lives.

Plants love Restore™

Not unexpected, but still fascinating to watch, Restore™ actually makes plants grow quite a bit faster. When you feed Restore™ to plants, their growth accelerates. Indeed, this is exactly what we would expect, considering the raw material in Restore™ is the very same medium of bacterial communication that helped ancient plants grow abundantly once-upon-a-time.

Stable redox molecules in supplement form are a major leap forward in human health

Restore™ rejuvenates gut health

Using Restore™, you can keep every membrane in your body perfectly protected from tight junction injury. This seals your gut lining and helps it heal. You can keep toxins out, and make toxins leave. You can improve hydration and nutrient absorption. And you can reduce the chronic inflammation that leads to dysfunction and disease.

Simply put, when you have strength of diversity in your microbiome, a lot of health problems go away, and are replaced with excellent mental and physical fortitude.

Of course, you still need to eat well

Dr. Bush and those involved in redox science are quick to point out that having immaculate tight junction protection is not a license to eat horribly. You can't just eat anything you want and hope to stay in excellent health for very long.

You still need to eat nutritious food, and avoid exposure to injurious substances, in order to keep your rate of healing up, and your rate of injury down.

The field of redox science has only just begun

The last 15-20 years of research into redox reactions have propelled medical science into new paradigms of healing, repair, and optimal aging. As a result, DNA and genetics are now believed to play a much smaller role in those processes than previously thought.

Over the last 4-8 years, medical scientists Gary Samuelson, PhD, and Dr. Zach Bush have successfully learned how to stabilize redox molecules and present them as supplements to the universal benefit of many thousands now taking Restore™.

The field has learned a great deal in the last 15 years. Yet everyone agrees, the majority of discoveries in the field of redox reactions have yet to be made. This is only the tip of the iceberg in understanding how redox molecules work, and in harnessing their ability to defend, heal, balance, and power our bodies.

The implications of redox science are staggering, as we can begin to see from a new discovery:

Scientists recently discovered something new about the gut that amazed, but didn't surprise, them

They found out that neural dendrites, which are remote extensions of the brain, actually poke through the lining of the gut, and hang out amongst the bacteria in the microbiome, without a membrane between them.

Presumably the brain and nervous system are literally reaching out into gut's ecosystem with nerve sensors to gather information about what's happening in the GI tract. **So it's possible that carbon redox molecules convey messages to the brain through these neural dendrites about the gut's inhabitants, its contents, and the condition of its cells.**

That's likely to be one of the reasons Dr. Zach Bush is seeing such rapid improvements in brain function when his redox molecule supplement Restore™ seals up the gut wall and rebalances the microbiome – including improvements in concentration, sleep quality, memory, and brain fog when taking Restore™.

Parting words

Some people have the good fortune to be gifted excellent health throughout their life and don't have to work very hard for it. Unfortunately, those people are quickly becoming an endangered species. The majority of the population will now have to fight at some point in their lives for the right to be healthy and live the life they desire (and deserve).

But although the health challenges facing us all may be daunting, or at least seem that way, be assured there are solutions to fix them all. The integrative/alternative community knows what causes ADD, GAPS conditions, and the chronic, degenerative diseases we all know by name.

And so the principles and practices you need to reverse a condition, as well as prevent it – those roads are now well-traveled by many practitioners and patients that have dealt with the same issues you or your loved ones are dealing with now. You're not alone.

The solution may not be as quick, and cheap, and easy as we would like them to be. But it's almost always far quicker, cheaper, and easier to fix health problems than it was to create them. So keep your expectations realistic short-term, and your outlook on health and life optimistic long-term.

In other words, don't accept anything less than excellent health, a fully-functional brain, and robust energy in your life, because you now have a choice. It is with these thoughts in mind that this collection of cutting edge principles and practices is offered to you, your family, and your loved ones.

The End (Part 2)

WHAT TO DO NOW

1. Help spread the word

If you found this information valuable to your health and life(style), I encourage you to go to Amazon.com and leave an honest review. Total number of reviews helps Amazon and prospective readers determine a book's worth – even more so than its average rating.

2. Visit the *Gut-Brain Secrets* website to learn more about healing your GUT, your BRAIN, and your BODY

www.GutBrainSecrets.com

3. Recommended reading

If you haven't read these books yet, I highly recommend you check them out. They're excellent.

- *Gut and Psychology Syndrome*, by Dr. Natasha Campbell-McBride.
- *Change Your Brain, Change Your Life*, by Dr. Daniel Amen.
- *Healing ADD*, by Dr. Daniel Amen.
- *Brain Maker*, by Dr. David Perlmutter.
- *The Body Ecology Diet*, by Donna Gates.
- *Driven to Distraction*, by Dr. Edward Hallowell.

Made in the USA
Columbia, SC
05 January 2021

30296835R00078